Becoming an Agent-Based Modeller

AF905611

Becoming an Agent-based Modeller takes you on a journey, from curiosity about social phenomena to generating them with a computer simulation. The book introduces agent-based modelling as a method to understand societies as complex systems, systems of interconnected, interdependent, and interacting agents.

The book is structured like a course rather than a textbook or reference book. The core of the book is an extended tutorial introducing the NetLogo programming language. Using the example of an epidemic, it explores how an agent-based model of social influence, situational awareness, and contextual decisions by individual agents about protective behaviour can help us understand plausible epidemic trajectories. The tutorial demonstrates how to build an agent-based model from scratch and use it in research, including the conceptualisation stage, the operationalisation and implementation, experimentation, and interpretation. This practical part of the book is embedded in chapters on theory, epistemology, and ethics of agent-based modelling.

With pedagogical tools including key discussion points, illustrations, and highlighted concepts, this is an ideal resource for courses on ABM, and for postgraduate students and researchers in the social sciences and beyond, who wish to develop their understanding of agent-based modelling.

Jennifer Badham is Assistant Professor in Social Data Science in the Department of Sociology at Durham University, UK. She is a computational social scientist, interested particularly in how social structure shapes the transmission of ideas, disease, or behaviour.

Corinna Elsenbroich is Reader in Computational Modelling at the School of Health and Wellbeing at the University of Glasgow, UK. She is a sociologist with a background in philosophy of science, sociology, and complexity social science methods.

Becoming an Agent-Based Modeller

Theory and Practice of Building Agent-Based Models in the Social Sciences

Jennifer Badham and Corinna Elsenbroich

Routledge
Taylor & Francis Group
LONDON AND NEW YORK

Designed cover image: Getty Images

First published 2026
by Routledge
4 Park Square, Milton Park, Abingdon, Oxon OX14 4RN

and by Routledge 605 Third Avenue, New York, NY 10158

Routledge is an imprint of the Taylor & Francis Group, an informa business

© 2026 Jennifer Badham and Corinna Elsenbroich

The right of Jennifer Badham and Corinna Elsenbroich to be identified as authors of this work has been asserted in accordance with sections 77 and 78 of the Copyright, Designs and Patents Act 1988.

All rights reserved. No part of this book may be reprinted or reproduced or utilised in any form or by any electronic, mechanical, or other means, now known or hereafter invented, including photocopying and recording, or in any information storage or retrieval system, without permission in writing from the publishers.

Trademark notice: Product or corporate names may be trademarks or registered trademarks, and are used only for identification and explanation without intent to infringe.

British Library Cataloguing-in-Publication Data
A catalogue record for this book is available from the British Library

ISBN: 978-1-032-61560-8 (hbk)
ISBN: 978-1-032-61558-5 (pbk)
ISBN: 978-1-032-61562-2 (ebk)

DOI: 10.4324/9781032615622

Typeset in Latin Modern Sans
by Apex CoVantage, LLC

Contents

Figures

Code Snippets for Tutorial Models

Tables

Preface

This book is the result of over a decade of collaboration between us, developing and delivering a course called Agent-based Modelling for the Social Scientist. The idea was to support a 'learn by doing' approach, to move participants through the two-fold change of perspective required to build their own models: agent-centric thinking, and computer programming. Programming itself has many strands, including explicitly and specifically designing what the program is supposed to do (which overlaps with agent-centric thinking), good programming practices such as continuous verification, and the keywords and syntax of the programming language used. Those three programming strands were woven into the *NetLogo* tutorial.

Jen then went ahead and in a (long and drawn-out) moment of madness wrote up the whole tutorial as an online resource for people to learn agent-based modelling by themselves. But whilst online resources are excellent, they often get less traffic than they deserve, and at some point we decided that it would be good to use the tutorial as the foundation for a textbook on agent-based modelling, also integrating the other parts of our course.

The tutorial material is presented in the same way as for one of the courses. Topics are introduced in the order in which they are needed for the model. New concepts are described immediately before being included in the model, then coded into the model to provide a concrete example. The reader is expected to run the model at multiple points as the code is developed. Such a structure leads to a certain amount of repetition. It also means that topics that may be introduced together in other books due to their similarity are instead introduced far apart.

The payoff of this structure is that the novice reader is able to experience the process of building a NetLogo model in the same way as they would build their own model: incrementally. The cost is that the book is much less useful for more advanced agent-based modellers. For example, it presents only limited theory. There is no intention to provide a complete guide or even meet the lower standard of a comprehensive and balanced introduction. Instead, the goal is to allow the novice to develop the skills necessary to navigate their own model development and access more advanced and specific materials as required. We hope that the book helps those novice modellers who are working away in their subject matter discipline, trying to develop expertise in agent-based modelling.

For the textbook, we added some of the theory missing from the online tutorial. The topics are mostly taken from the course but with a little more detail and pointers to other material you can use to deepen your understanding. We also added a more detailed discussion on ethics and some additional reflections on the epistemology of agent-based modelling.

Building an agent-based model is not just a technical process of writing code that follows syntax and logic rules; it requires a specific way of thinking about the system you are representing. Adopting that perspective is central to becoming an agent-based modeller. This way of thinking is also captured in our choice of cover image. Origami has you create representations of real world objects. You follow a process of paper folding to make that representation. It is also a nod to James M. Sakoda whose work spans agent-based modelling and the art of origami.

Acknowledgements

We would like to thank first and foremost our roughly 150 students who have suffered through the course with us over the last ten years. It's a tough three days, in particular for those who have no programming experience at all. We enjoyed every single iteration, had great feedback along the way, and some of you have kept in touch and continued in the field of agent-based modelling.

We would also like to thank some of our colleagues who contributed to the early iterations of the course: Tina Balke, Peter Barbrook-Johnson, Nigel Gilbert, Kavin Narasimhan, and Nicolas Payette. At the time all of us were working at the Centre for Research in Social Simulation (CRESS) at Surrey with Nigel Gilbert.

The book centres around the tutorial, which has been available online for a few years now. We would like to thank Jamal Rahmani, Liz Groff, Jeon-Young Kang, Hal Kohn, and Pietro Terna for testing and improving the online tutorial, and Simran Garg for doing a test run of the book version. We would like to thank Duncan Stone for braving a test read of the book to check for consistency, lucidity, and to gauge the boredom factor for a non-modeller. He got to Chapter 3.

Part I

Conceptualising Agent-based Models

1

Modelling the Social World

This chapter discusses the concept of societies as complex systems, emphasising the importance of heterogeneity, interaction, non-linearity, and emergence. It introduces the idea of agent-based modelling, a computational methodology that allows researchers to create, analyse, and experiment with models composed of agents that interact within an environment. The chapter will introduce some classic models, such as the Schelling model, a Game Theory model, the Sugarscape model, and an Opinion Dynamics model, to illustrate how agent-based models can display complex behaviours and answer fundamental questions about social systems.

1.1 How to use this book

Welcome to the journey of Becoming an Agent-based Modeller. This book will be your guide traversing the stormy seas of simulation modelling for the social sciences. It will help you to conceptualise, design, implement, interpret, and communicate your model. Throughout the book, we will try to keep all of these activities in mind and make clear how they interdepend. You cannot implement a model that is badly designed and you cannot clearly communicate a model that is poorly conceptualised.

The first three chapters of the book discuss some general theory, concepts, and ethics of agent-based modelling. The heart of the book is the tutorial in Part II. It will guide you step-by-step through the process of building an agent-based model, including conceptualisation, operationalisation, and coding. There is a chance that in the tutorial you might feel a little lost at times, particularly if you are completely new to coding and you do this as a self-learner. We can only encourage you to persist. Often things will become clearer after doing the next section in the tutorial. Part III discusses other topics you need to become an agent-based modeller. The first two chapters discuss aspects of building a model not covered by the tutorial: movement and GIS (Chapter 11), and additional good programming practices (Chapter 12). This is followed by chapters about using your model (Chapter 13) and how to relate the model back to the real world to ensure that it is adequate for such use (Chapter 14).

The tutorial model concerns protective behaviour (such as good hand hygiene) when faced with an epidemic. It is unlikely that the model that you want to build is about the same topic. However, regardless of your topic, the core process being simulated is common to many agent-based models: people making decisions based on their own motivations and circumstances and those decisions in turn affecting the world around them.

At the end of the book, you will have become an agent-based modeller. Strap yourself in for the journey!

1.2 Understanding Societies

In an influential interview in *Woman's Own* magazine Margaret Thatcher, UK Prime Minister (1979-1990), said the now often-quoted sentence "There is no such thing as society". She follows on to say that "There are individual men and women and there are families and no government can do anything except through people and people look to themselves first."

There is a strange paradox in these sentences. The intention was to assert a strong form of individualism, as government cannot do anything except through people, and to assert a strong form of selfishness as people look to themselves first. However, if we look closer, Thatcher starts with an existence claim of three categories, 'men, women, and families'. The existence of families means there is something 'over and above' the individual, a collective entity with a status equal to the individual. But what is a family? Does she mean a nuclear family of parents and children? Can this hold for multi-generational families? What about extended families including in-laws, aunts and uncles? And what about families of choice, communes, friendship circles, or religious congregations?

Once we open Pandora's box of asserting the existence of more than just the individual, a host of questions arise about *what* we are investigating in the social sciences and *how* we can find out about it. What we are investigating is a question of *ontology*.

Ontology

Ontology is the study of what there is, what exists.

In the social sciences there are two extreme ontological positions: individualism and holism, the former stating that only individuals exist, the latter that there exist social structures over and above individuals. The dispute between individualism and holism (or structuralism) is a longstanding debate in the philosophy of the social sciences (Kincaid, 2014). Within the social sciences themselves the debate is usually couched as the concepts of agency and structure. If we are assuming the existence (and causal relevance) of social structures, individuals quickly become mere puppets without autonomy and without the capacity to make autonomous decisions (Zahle, 2023). On the other hand we have agency of individuals, preserving autonomy but questioning the existence of social structures as real entities with causal powers. Ontologies of course do not have to be as extreme as individualism and holism; intermediate positions assert both individual agency and structure, e.g. Giddens' structuration theory (Giddens, 1984).

One important position in the social sciences is Methodological Individualism. Methodological Individualism explains social phenomena primarily through the actions, motivations, and interactions of individual agents but allows for social structures resulting as aggregates. It holds that macro-level outcomes emerge from the interactions of individuals, but once they have emerged they can be essential to understand personal beliefs, choices, and incentives to explain broader societal trends (Elster, 1989).

Ontological questions of what exists are not mere abstract discussions; they have consequences for our *epistemology*.

Epistemology

Epistemology is the study of what we can know and how we can know it.

Epistemology in the social sciences is the branch of philosophy that examines the nature, sources, and limits of knowledge about societies and social behaviour. It explores how we come to understand social phenomena, the assumptions underlying various research approaches, and the criteria by which social knowledge is validated. Central debates in this area include the tension between positivist approaches, which advocate for objective, quantifiable methods to uncover social laws, and interpretivist perspectives that emphasise subjective experience, meaning, and the social construction of reality. Additionally, epistemological inquiry in the social sciences considers the influence of context, power relations, and cultural norms on both the production of knowledge and the research process itself, raising questions about the possibility of truly unbiased, value-free research in the study of human behaviour and institutions.

Exploring this Pandora's box is the bread and butter of the social sciences. Whilst in the natural sciences there is often an overarching paradigm and sequential paradigm shifts (Kuhn, 1970), in the social sciences many paradigms exist in parallel, gaining and losing importance or finding synergies over time. This plurality of theories in the social sciences is mirrored by a plurality of methods and disagreement about methodology and which method(s) best capture the social world.

Methodology

Methodology refers to the systematic approach or framework used in a particular field of study to guide how research is conducted. It involves the principles, rules, and procedures that researchers follow to collect, analyse, and interpret data.

The methodological divide in the social sciences can be roughly described as the divide between quantitative methods, focusing on measurement, generalisation, and objectivity, and qualitative methods, focusing on narratives, individuation, and subjective experiences. These two groups of methods rest on distinct epistemological and ontological assumptions. Qualitative approaches generally embrace an interpretivist epistemology, viewing knowledge as co-constructed through social interactions and subjective experiences. Ontologically, qualitative researchers see social reality as fluid, context-dependent, and socially constructed. In contrast, quantitative methods are rooted in positivism, holding that objective, measurable reality exists independently of human perceptions. This approach assumes that social phenomena can be quantified, generalised, and predicted based on stable causal laws. In recent years there has been a recognition that these methodologies offer complementary insights. While qualitative methods provide deep, context-rich understandings, quantitative methods aim for objectivity and replicability, and both of these are needed to understand the social world.

We have already introduced Methodological Individualism, which focuses on understanding social phenomena by analysing the actions and interactions of individuals. It is an important position for understanding agent-based modelling, which can be seen as an operationalisation of this approach by simulating individual agents with specific behaviours and decision-making processes. Through these simulations, researchers can observe how simple rules at the individual level lead to complex collective outcomes, thereby providing valuable insights into the dynamics of social systems. The emergence of social structures, proposed in Methodological Individualism as the mechanism underlying the social world, is directly replicated in agent-based models (Manzo, 2023).

The ontological starting point of this book is that societies are complex systems; systems of interacting components. Note that these components do not need to be individuals. The epistemological starting point is one of structural realism, that we can know things about system components as well as whole system properties. The methodological starting point is that agent-based modelling is a methodology that helps us to understand societies as complex systems by mirroring the processes underlying social phenomena in a computer program.

In the following sections we will dive deeper into complex systems, define an agent-based model, and look at some classic agent-based models and what they can tell us about the social world.

1.3 Complex Systems

There is no agreed definition of a complex system. At a minimum, a complex system is a system of interacting components. This, however, is not a sufficient description of complex systems as there are systems of interacting components that are not complex (but may be merely complicated, think of a car engine).

Often complex systems are captured best by describing particular characteristics of the system that make it complex. While these descriptions emphasise different aspects, the key to complexity is how the parts of the system interact and what overall system behaviours *emerge* from these interactions. If the behaviour of the system cannot be understood by examining the parts separately, and arises instead from the interaction between parts, the system is exhibiting complex behaviour. That is, a complex system is more than just a complicated system with many parts that are related in some way; the interaction between the parts must be critical in generating the behaviour of interest (Miller and Page, 2007).

Consider a traffic jam. This phenomenon is generated not just because there are many moving cars, but because of the way in which interactions between the drivers of those cars introduce cascades of braking and accelerating. A single blockage, such as a set of traffic lights, braking to avoid an animal on the road, or correcting a drift too close to the car in front, can trigger a traffic jam. Differences in drivers' preferred speeds, separation, and reaction speeds can amplify any slight variations. Analysing individual cars would not provide any understanding of the phenomenon of a traffic jam. In different circumstances, such as moving smoothly on a highway, the same cars can be considered somewhat independently as their behaviour does not significantly affect the behaviour of others.

The modern term of complexity was introduced to describe a set of "problems which involve dealing simultaneously with a sizable number of factors which are interrelated into an organic whole" (Weaver, 1948, p. 539). However, the essence of the idea was encapsulated much earlier by Aristotle in the phrase "the whole is one, not in the manner of a heap but of a syllable" (Aristotle, 2011, 1041b, more familiar as "the whole is other than the sum of its parts").

Weaver (1948) recognised that complex problems represented a relatively unexplored area in science and were particularly prevalent in biological, medical, psychological, economic, and political sciences. A modern commentator would add social and environmental sciences, at least, to this list. Complex systems questions include some of the most pressing policy issues: how to limit the extremes of economic cycles, what species are critical to maintaining the food web, where is the point at which we cannot recover from climate change, how to avoid an influenza pandemic. All of these questions involve phenomena arising from behaviours and interactions of individuals. For example, influenza is transmitted between people, which can only occur through contact between infected and susceptible people. Economic cycles are the consequence of millions of transactions where institutions, firms, and individuals purchase goods and services from other institutions, firms, and entities.

While economies, food webs, and epidemics may be referred to as complex systems, this language is somewhat misleading. What is of interest is a particular behaviour of the system, not the obvious entities within the system. To bridge this gap, the term 'system' must be understood in its widest sense, so that parts of the system are not just the individual components, but also the rules of interaction. To return to the traffic jam example, the system does not consist only of cars, drivers, and roads, but must also include the characteristics and behaviour of the drivers that contribute to the traffic jam. To make this comprehensive scope of 'system' clearer, this book uses a definition of complex systems that explicitly focuses on behaviour.

Complex System

A system is complex if the system's behaviour is the *emergent* result of the *interactions* between *components* of the system.

Importantly, system components do not need to be uniform but complex systems can consist of many different, or *heterogeneous*, parts. While heterogeneity, interaction, and emergence are often seen as the fundamental ingredients of complex systems, there are additional features of complex systems that are important to understand for the rest of the book.

The first important additional attribute of complex systems is that they are *open systems*. There is no natural 'boundary' around the system. Even for the traffic jam discussed earlier, in some cases the weather is best considered part of the complex system (e.g. a hailstorm), whereas other times it is movement patterns of the wildlife in the forest that is needed to understand the system. It is important to understand that the whole world is the complex system we are faced with and we are making a decision of how to carve it up into complex subsystems that we try to understand in more detail. Importantly, every subsystem will be situated within a bigger system, a property of complex systems often referred to as 'nested complex systems'. This will become an important aspect of complex systems when decisions need to be made about what is modelled.

The second important additional attribute of complex systems is that they *change over time*. A lot of social science is quite static. We try to understand the relationships in the social world analysing snapshots at certain time points (e.g. survey data). Complexity necessitates dynamics, movement, and change. The reason for this is that a complex system is defined through the interactions of its constituent parts and an interaction is a process in time. By default this means that static methods will not suffice to understand complex systems.

The third important additional attribute to discuss in the context of modelling complex systems is that they function on *multiple scales and levels*. These multiple scales and levels capture that changes on the microlevel lead to emergent changes at more aggregate levels; but it also captures an aspect of complex systems that the actions and interactions themselves operate across levels, meaning complex systems need to be studied from different perspectives and experiences.

The fourth important additional attribute is *feedback*. Causal feedback loops are one of the most defining features of complex systems. It means that a consequence B of cause A can again cause changes in A in a future time-step. Feedback is often connected to emergence. Think for example about a situation in which more and more people buy a certain phone, which then becomes a fashion item, which in turn leads to more and more people buying the phone.

Each of these features poses a challenge to our understanding of a complex system. A particular challenge is posed by the impossibility of deconstructing the system into its component parts as it is their interaction, across *multiple scales and levels*, within an environment (*open system*), over time that produces the complex behaviours. So we need to think about methods that allow us to honour these features.

As well as identifying the class of complex problems, Weaver (1948) also suggested that progress could only be made by multidisciplinary teams developing new computer-assisted methods. Agent-based modelling is one such method.

1.4 Models and Simulations

The recognition of complexity has some important consequences for how we can go about understanding what is 'going on'. Experimental approaches are difficult in complex systems. Experiments, by design, rely on the possibility of isolation and control (e.g. in a laboratory). As complex systems are defined through the interaction between components, system behaviour cannot be understood by isolating system components. Once isolated, the components will most likely behave differently, making inference from experimental outcomes to real world outcomes difficult if not impossible (Cartwright and Hardie, 2012).

But it is also impossible to understand the whole world in its complexity. The way to go about understanding complex systems is to investigate subsystems. Whilst we cannot isolate system components in the hope of understanding their behaviour, we can identify subsystems to investigate, and this investigation is best done by building a model of the subsystems.

So, what is a 'model'? Often the first thing one thinks about when hearing the term model is a model aeroplane, a model car, a fashion model. These are things or people standing in for the 'real' things, the real flying aeroplane, the real driving car, the real clothes-wearing person. They are not, however, exact replicas. The model aeroplane usually does not fly, and even if it does it will not hold passengers (if it had all the features of the real aeroplane it would be the plane itself, not the model!). The model car may look like a car, but it cannot be used to travel. And most people who will eventually wear the clothes being modelled have different body shapes than most fashion models. Ultimately, the choice of what to include in the model, and at what level of detail, is guided by the purpose of the model. For example, the lack of an engine is immaterial in a collectible model car; the goal is a faithful representation of the body shape and styling. On the other hand, a racing car simulator needs to replicate the car's driving behaviour as closely as possible but, as it is being simulated in a computer, the car has none of the physical and material components of the real car.

A second kind of model that is familiar to many people is a mathematical representation of a system or state of affairs. This includes dynamical systems, statistical models, differential equations, and game theoretic models.

Both the material and mathematical models represent something concrete, a target system. These ideas define what we mean by 'model' (Hartmann, 1996; Greenberger et al., 1976).

Model

A model is a representation of relevant features and relationships of some target system.

This definition has three elements; let's start from the back of the sentence.

1. *Target system*: the model is 'of' something, a concrete or specific 'thing';
2. *Representation*: the model is not the thing itself;
3. *Relevant features and relationships*: a model only represents some aspects, not all.

In this book we are concerned with a particular subgroup of models, agent-based models, which are a subgroup of simulation models. Hartmann (1996) states that a simulation "allows scientists to imitate one process by another process". We stick to the terminology of *representation* for the purpose of this book.

Simulation

A simulation is the representation of a process by another process.

Simulations are essential for understanding complexity due to the importance of change over time of complex systems; we need to understand the process of change.

There are at least two ways to think about a process, and therefore to represent it. One way is to describe the process from the perspective of an external observer. Hartmann's (1996) definition of process as simply a description of the "temporal sequence of states of a system" captures this external observer perspective. Alternatively, the process can be described from the perspective of the interacting parts of a system, what actions they take and why. Returning to the traffic jam example, an observer may describe it as a line of vehicles that are moving slowly (or stalled) over a distance of several hundred metres. The drivers, however, are within the traffic jam and do not know its length or other characteristics. Instead, they perceive only the local traffic, and respond to the vehicle in front of them by starting and stopping as appropriate. Some drivers may look further ahead, but any information gained has little influence on their progress.

Simulations, just like static models, can be physical representations; just think of a motorised simulation model of the solar system which represents the planets' orbits around the sun. In this book we are concerned with computer simulations, using computer programs to replicate processes found in the real world.

There are different computational simulation methods in the social sciences, taking different approaches to the question of what kind of process is modelled. This book is about agent-based modelling, and in the next section we will look in more detail at why agent-based modelling is a particularly good simulation method for complex systems. The other prominent simulation methods are microsimulation, discrete event simulation, and system dynamics, each of which represents some aspects of complexity.

1.4.1 Microsimulation

Microsimulation goes back to an idea of Guy Orcutt (1957). Orcutt's approach uses individual-level data to model economic and social outcomes. A key feature of this method is that all modelling is conducted at the individual level, enabling the analyst to assess how external factors, such as government policies, affect individuals, families, households, businesses, and more. These individual impacts can then be aggregated to reveal the effects on various population sub-groups (e.g., by age, income bracket, etc.).

Microsimulation uses detailed, individual-level data and applies transition probabilities to simulate how these entities change over time, like tax and benefit changes (Richiardi et al., 2021), demographic developments, pensions (Salonen et al., 2021), or demands on the education system (Hannappel et al., 2012).

The ability of microsimulation to model sub-groups within a population makes it a powerful tool for understanding distributional impacts, or how different groups are affected by changes. It is widely used in fields like economics, public policy, healthcare, and social science to analyse the distributional impacts of policy reforms, taxes, welfare programmes, or demographic changes.

Key characteristics of microsimulation:

1. Individual-Level Analysis: Focuses on simulating change at the micro level, using data about individuals, households, or firms.
2. Data Driven: Microsimulation applies transition probabilities elicited from past data to model state changes for individuals.
3. Aggregation of Results: After simulating individual changes, results can be aggregated to reveal broader trends or impacts across different subpopulations.
4. Dynamic or Static: Microsimulation can be static, assessing change between two policy options (a 'morning after' scenario), or dynamic, where individual characteristics evolve over a longer time (e.g., as people age, their income or health status changes).

It is an excellent simulation method to model heterogeneity and (limited) change over time, two important aspects of complexity. Importantly, however, in a microsimulation the individuals (or system components) do not interact with each other, meaning microsimulations do not directly model complex systems.

1.4.2 Discrete Event Simulation

Discrete event simulation is a modelling technique used to simulate the operation of systems in which changes, or "events", occur at specific, discrete points in time. Processes that have a duration are modelled by events at the start and end of that process. It is commonly used to model systems such as manufacturing processes, supply chains, or healthcare operations, where the state of the system changes only when certain events take place.

In a discrete event simulation model, time progresses in jumps, moving from one event to the next, rather than mirroring minutes, days, or weeks. Each event causes a change in the state of the system, and these events are typically triggered by certain conditions or actions, such as the arrival of a customer in a queue, the completion of a task, or the breakdown of a machine. Between these events, the system is assumed to remain in a steady state.

Key characteristics of discrete event simulation:

1. Discrete Time Progression: Time advances from one event to the next, not continuously.
2. Event-Driven: Changes in the system occur only when specific events happen.
3. State Changes: The system's state is updated at each event, reflecting the impact of that event on the system.

Discrete Event Simulation is used in various fields to evaluate the efficiency of processes, optimise resource allocation, and predict system performance under different conditions.

1.4.3 System Dynamics

System dynamics modelling is an approach used to understand and simulate the overall behaviour of complex systems over time. It focuses on the interrelationships between different elements of a system and how these relationships generate feedback loops that influence the system's overall behaviour. By representing systems as collections of interconnected stocks (populations and resources), flows (rates of change), feedback loops, and delays, system dynamics helps analyse the long-term impact of changes and identify potential unintended consequences.

System components are represented at a high level of aggregation. That is, stocks count some resource but do not represent individual elements, treating them as interchangeable. For example, an infectious disease model would represent the susceptible, infected, and recovered populations as separate stocks and model the progression of the disease at the population level through infection and recovery rates (flows) that depend on the size of those stocks.

Key components of system dynamics include:

1. Stocks: These represent the accumulation of resources or quantities within the system (e.g., population, capital, inventory).
2. Flows: Flows are the rates at which stocks change over time (e.g., birth rates, production rates, or consumption rates). They describe the movement of resources into and out of stocks.
3. Feedback Loops: Feedback loops describe how changes in one part of the system influence other parts, creating either reinforcing or balancing effects.
   (a) Reinforcing (Positive) Loops: These loops amplify changes, leading to exponential growth or decline (e.g., population growth where more people lead to more births, creating a cycle of expansion).
   (b) Balancing (Negative) Loops: These loops counteract changes, pushing the system toward equilibrium (e.g., when rising inventory levels trigger lower production rates).

4. Delays: Time lags between actions and their effects are captured, showing how certain reactions or responses take time to materialise in the system.

System dynamics is often used to model complex systems in areas like economics, environmental science, public health, and organisational management. Examples of use include understanding population growth, managing supply chains, analysing the spread of diseases, and evaluating the long-term sustainability of environmental policies. The visual representation of these models often uses causal loop diagrams and stock-and-flow diagrams to illustrate relationships and feedback mechanisms within the system. It helps understand interactions of system components, feedback and change over time. As it is modelling at the aggregate level, system dynamics does not focus on heterogeneity.

Each of these simulation methods models some aspects of complexity very well. Microsimulation is excellent at modelling heterogeneity, i.e. differences between individuals to better understand differential outcomes across subpopulations. Discrete Event Simulation allows for the better understanding of time sensitivity of events and bottlenecks in a system. Systems Dynamics is excellent at modelling feedback and cumulative processes. But they all describe the process of change rather than generate change.

Earlier, we said there are two ways of understanding the process of change, the outside observer view and the participant view. For the traffic jam example the participant view means to not look at the line of cars moving slowly along the road but to simulate individual drivers, each interpreting their environment (distance to the car in front, speed of the car in front), and combining that with their own characteristics (preferred speed, current speed) to determine an action (accelerate, brake). By directly representing the drivers at the individual level, their interactions and decisions generate the processes of the overall complex system, bringing about emergent patterns at the aggregate level.

We turn to this participant view of change in the next section where we introduce agent-based modelling.

1.4.4 Agent-based Modelling

An agent-based model is a computational simulation model of autonomous agents interacting with each other and an environment. Already from this definition it becomes clear why agent-based models are particularly suitable for modelling complex systems: they directly replicate the ingredients of complex systems, i.e. a system of interacting parts (called *agents*) (Gilbert, 2008).

Key components of agent-based modelling include:

1. Micro-Representation: System components are represented at a low level of aggregation in the form of individual agents.

2. Autonomy: Agents are autonomous and active system components, able to make their own decisions and affect the system.

3. Interaction: Agents can interact with each other and their environment.

4. Emergence: Micro-behaviours and -interactions lead to macro-level patterns.

By directly replicating the agents and their interactions, agent-based models can replicate many features of a complex system. This includes features also captured in other simulation methods, such as heterogeneity (microsimulation), event-driven systems change (discrete event simulation), and feedback (system dynamics). In addition agent-based modelling can replicate other features of complex systems, such as interactions at multiple levels and scales (micro-macro interactions) and emergence.

Understanding the connection between individual (micro-level) behaviour mechanisms and overall (macro-level) system behaviour is an important use of agent-based models. Specifying the system at the micro-level, including the ability of micro-components to act and interact, necessitates an *agent-centric perspective*. Such a perspective allows the interactions that drive the behaviour of a complex system to be directly represented in a natural way. It is particularly valuable where interactions are combined with heterogeneity because the methodology enables different agents in the same situations to take different actions, and the same agent in different situations to take different actions. Some of the richness of the social world can therefore be represented in an agent-based model.

1.5 Four Classic Model Types

We have stated that agent-based modelling is a powerful tool for understanding complex systems, especially in the social sciences. The first question to ask when using this method is, why? Why would we choose agent-based modelling over other methods, techniques, or tools? The reason is that agent-based models can mimic the processes underlying social phenomena.

An agent-based model is a computational method used to create, analyse, and experiment with systems composed of agents that autonomously interact within a defined environment (Gilbert, 2008). These agents can represent anything (people, companies, animals, or even particles) depending on the system being modelled. Their behaviour can be as simple as following others in a social network or more complex such as trading resources in an economic market or learning behaviours from each other.

In this section we look at four models that represent some human behaviours that can be seen as fundamental dynamics underlying social processes.

1.5.1 Segregation

The first dynamic segregation model focusing on the interactions of individuals was developed by James M. Sakoda (1949) in his PhD thesis and later published in Sakoda (1971). Sakoda explicitly used the language of *complexity* and focused on the necessity to understand *social interactional processes* (Hegselmann, 2017). Despite Sakoda's precedence by a couple of decades, the more familiar agent-based model of segregation is the one developed by economist Thomas Schelling (1971), trying to understand residential segregation in American cities.

Agents of two colours are placed on a grid and allowed to move if they are dissatisfied with the diversity of their neighbourhood. If an agent has more neighbours of the other colour than specified by their happiness threshold (their tolerance for the other colour), they move. In the original model that Schelling executed manually on a checkerboard, agents moved to the closest patch that satisfies their requirements. In what follows we will use a model where dissatisfied agents instead move to a random free space.

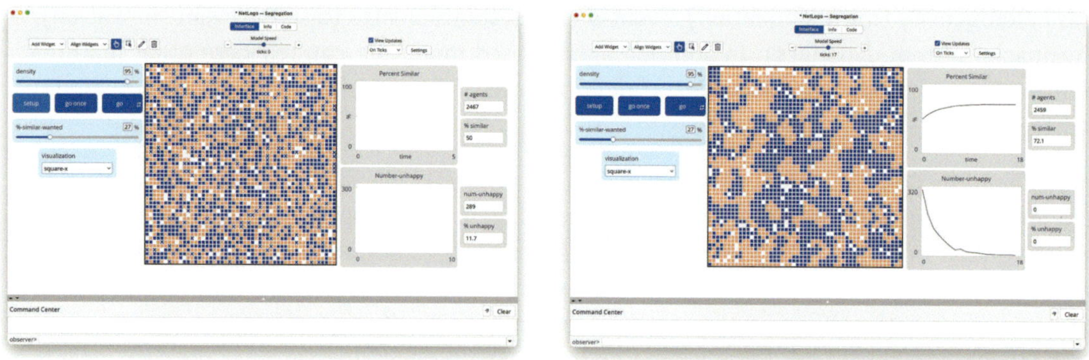

(a) Setup Configuration (b) Final Configuration

FIGURE 1.1: Segregation model inspired by Schelling's segregation model (Wilensky, 1997).

Figure 1.1a shows the setup configuration with agents randomly allocated across the grid, with 50% of each colour. Agents want 30% of their neighbours to be the same as themselves. As each agent has eight neighbours, this translates into at least three neighbours of the same colour. With this setup we reach segregation levels greater than 75%, as shown in Figure 1.1b. Even when agents have a mild preference for neighbours of their own type, the model shows that high levels of segregation emerge, demonstrating the powerful role that local preferences can play in shaping city-wide outcomes.

It is important to stress that this model does not suggest there is no racism underlying segregation. It does not suggest that people have a mild preference for similar neighbours. It shows that if we want to do something about segregation, we have to work very hard indeed as segregation is so easy to bring about *even if* people only have mild preferences. This model explores macro-phenomena resulting from agent movement.

What is important to note is the iterative nature of the model in which one agent's move changes the environment of other agents, making them move in the next step, leading to cascading movement throughout the system. While very simple, the dynamics of moving towards or moving away are fundamental to many social phenomena of group formation and segregation. For example, Stoica and Flache (2014) explore segregation in schools in the Netherlands building on the Schelling model and Castellani et al. (2015) uses movement dynamics to explore the persistence of place-based health inequalities.

1.5.2 Strategic Interaction

One early, and large, area of application of agent-based modelling has been evolutionary Game Theory. Game Theory is the mathematical study of strategic interaction between players. Game Theory assumes rational agents who choose actions to maximise their utility. For Game Theory, an agent A's utility is dependent on the actions of another agent B, and *vice versa*. This interdependence can lead to suboptimal equilibria.

A well-known Game Theory game is the Prisoner's Dilemma (Poundstone, 1993). It is a cooperation game with an asymmetric payoff matrix. The interesting feature of the Prisoner's Dilemma is that whilst both agents cooperating would yield the highest overall payoff ($3 + 3 = 6$ for the payoff matrix in the model in Figure 1.2), each individual has an incentive to defect as their

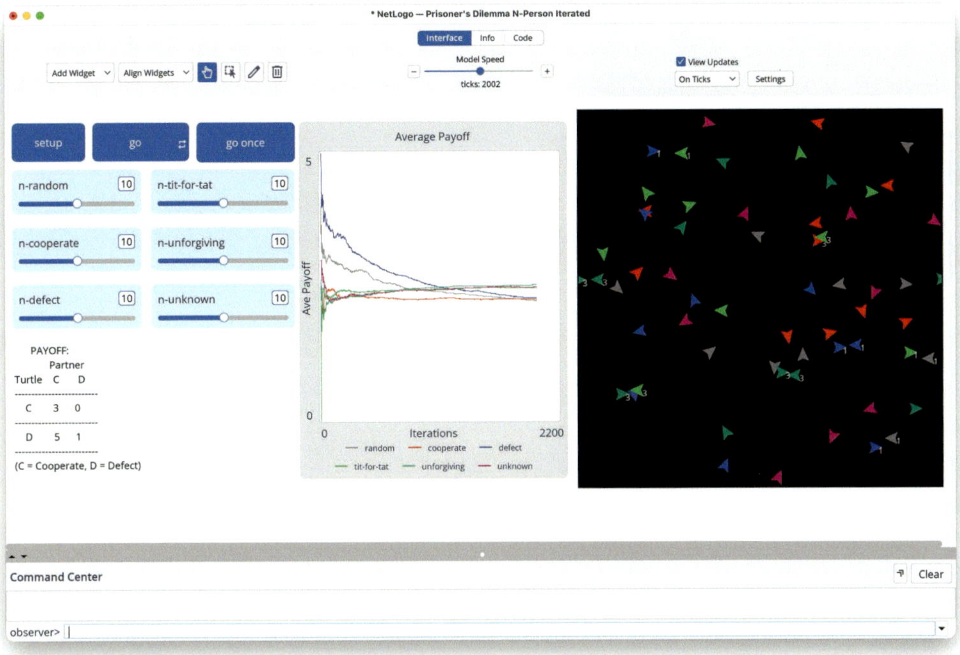

FIGURE 1.2: Iterated Prisoner's Dilemma: Different strategies competing. The defect strategy starts very successfully with the highest average payoff but is overtaken in the long term by the adaptive strategies of TIT-FOR-TAT and unforgiving (cooperate until defected against, then always defect)(Wilensky, 2002).

individual payoff will be higher regardless of the other player's action (5 versus 3 if the other agent cooperates, and 1 versus 0 if they defect). As both players have the same incentive, the equilibrium of the Prisoner's Dilemma is that both players defect, ending up in the worst possible collective payoff $(1 + 1 = 2)$.

Whilst in the single-shot Prisoner's Dilemma defection is indeed the dominant strategy, this changes under conditions of repeated interactions. These repeated interactions are investigated in the field of evolutionary game theory. Evolutionary game theory can be analysed mathematically, but there are limits to the mathematical treatment as certain population compositions become mathematically intractable. Here agent-based modelling can help by simulating the interactions, mutations, and selection mechanisms for strategies over time.

In 1980, Robert Axelrod (1980) held a tournament of various strategies for the Prisoner's Dilemma submitted by well-known game theorists. A *strategy* was a specification of whether to cooperate or defect based on the previous moves done and experienced by a player. A simple version of this tournament is shown in Fig 1.2, with six competing strategies.

Three strategies do not react to past experiences: always *defect*, always *cooperate* or play *randomly* (cooperating 50% of the time). *Unforgiving* and TIT-FOR-TAT adapt to past experiences, with the former cooperating until they experience their first defection, from then on they will always defect. On the first move TIT-FOR-TAT cooperates and then mirrors its opponent from the previous move. A sixth strategy, *unknown* is listed but is currently set to TIT-FOR-TAT. You can invent a strategy and try it out yourself!

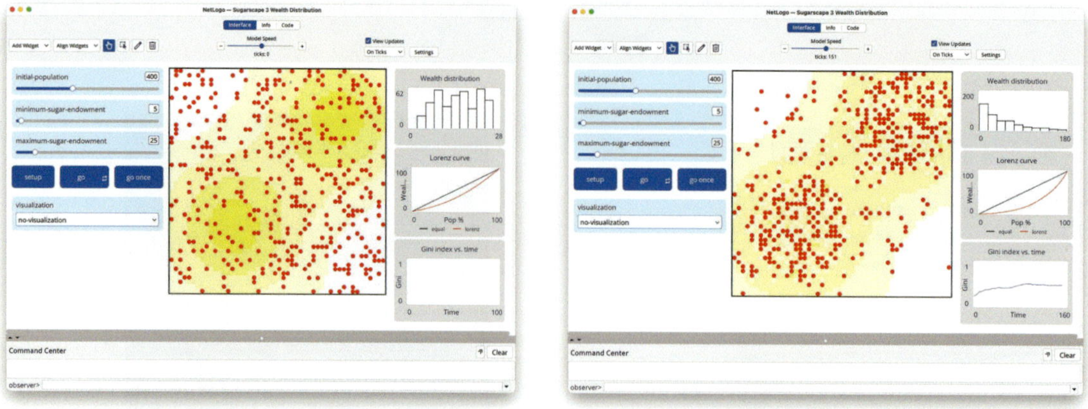

(a) Setup Configuration

(b) Wealth Accumulation

FIGURE 1.3: The Sugarscape Model of Wealth Distribution (Li and Wilensky, 2009).

TIT-FOR-TAT was the strategy that won the original tournament, by consistently outperforming other strategies. A population with TIT-FOR-TAT players will end up in cooperation as the mirroring allows players to exploit the higher cooperation payoffs without being 'suckered'. Simulation research allows a better understanding of how societal composition and adaptation to other people's behaviours results in cooperative or defecting societies.

Simulation becomes particularly interesting when game theory is implemented with evolutionary dynamics, where successful strategies have a higher number of offspring. Axelrod and Hamilton (1981) show how reciprocity can lead to cooperation, wins against a wide range of other strategies, and can resist invasion once established.

This line of evolutionary game theory research has been extensively pursued (Sandholm, 2020), with a range of games (Prisoner's Dilemma, Stag, Couples, etc.) and under increasingly complex conditions of punishment (Lettieri and Parisi, 2013), reputation (Grow and Flache, 2019; Giardini et al., 2013), or trust (Janssen et al., 2022; Klein and Marx, 2018). The point of agent-based modelling here is clear. We have a situation in which agents interact with each other with a clear payoff structure and a range of behaviour options; the simulation modelling helps us to understand the macro-patterns emerging from the iteration of interactions (Tesfatsion, 2023).

1.5.3 Wealth and Exchange

The third classic model we want to discuss is *Sugarscape*. Developed by Epstein and Axtell (1996, Ch 2), this model simulates a population consuming and exchanging resources from an environment. Agents (red dots in Figure 1.3) need to harvest and consume sugar to survive. They find sugar in the environment (the darker the shade the more sugar), which regrows after harvesting. Agents can store sugar, allowing for wealth accumulation. Starting from a normal a normal distribution of wealth and low inequality (Figure 1.3a), inequality increases (Gini index and Lorenz Curve) with a skewed wealth distribution after running the simulation (Figure 1.3b).

Once we have resources and agents, we can also introduce trade, emulating economic markets. In an extension to Sugarscape with two resources, agents can trade sugar and spice, according to their preferences (Epstein and Axtell, 1996, Ch 4). Sugarscape reveals how geography (such

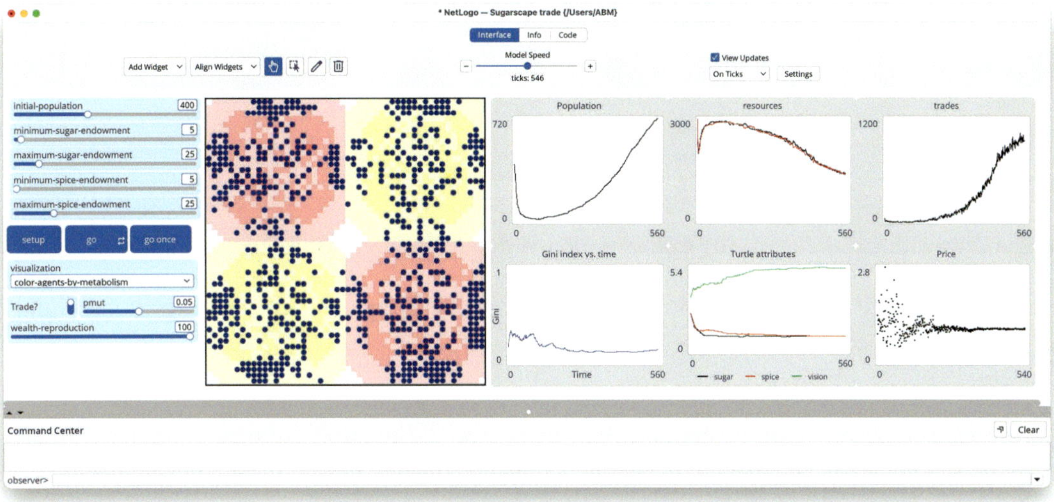

FIGURE 1.4: Agents trading sugar and spice (from Janssen, 2020)

as the spatial distribution of resources) affects market prices, showing greater price fluctuations and lower overall quantities traded than predicted by classical economic theory. Figure 1.4 shows the number of trades and the market price. Importantly, the emerging market price fluctuates a lot for a small population with relatively few trades, stabilising as the population and number of trades increases. This model highlights the importance of physical space and agent interaction in shaping market dynamics.

This model replicates the more general process of market exchange. Sugarscape implements a bilateral exchange of sugar and spice but it could be the exchange of goods for money or information. We could see an epidemic model as a limiting case of an exchange mechanism as a unilateral exchange of a virus. Important in Sugarscape is the geographical restriction of exchange, the availability and spatial distribution of resources and the number of people an agent can meet. It also allows for heterogeneity of agents, for example different preferences for sugar and spice.

The fundamental process of exchange has been extensively pursued within agent-based modelling. Hamill and Gilbert (2015) and Tesfatsion (2023) provide excellent overviews of agent-based modelling in economics.

1.5.4 Social Influence

The fourth social dynamic explored by classic agent-based models is that of social influence. People are influenced by what others do; we change our outfits in accordance with fashion, we adopt behaviours, like shaking hands or hugging as a greeting, and we change our opinions talking to other people. These dynamics can all be summed up as the dynamics of social influence.

One large field of research in agent-based modelling of social influence is opinion dynamics (Mastroeni et al., 2019). One of the earliest of these models is the Bounded Confidence model (Hegselmann and Krause, 2002). The Bounded Confidence model starts from the assumption that people align their opinions, but only if they are already sufficiently similar. Opinions in this

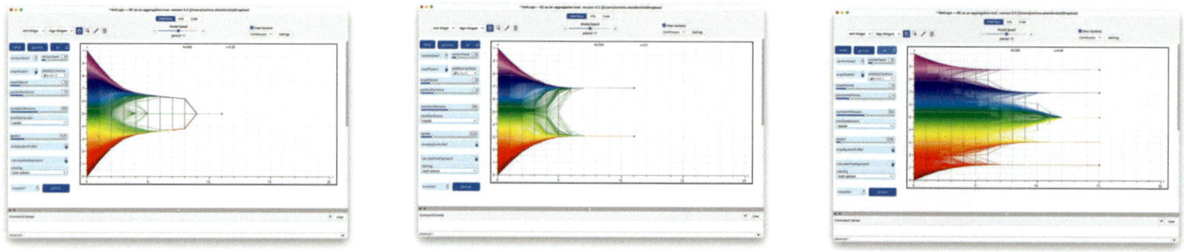

(a) Configuration with $\epsilon = 0.25$ (b) Configuration with $\epsilon = 0.2$ (c) Configuration with $\epsilon = 0.08$

FIGURE 1.5: Bounded Confidence Model (unpublished NetLogo version of Hegselmann and Krause, 2002, private communication)

model are operationalised as positions on a line from 0 to 1 (imagine politically left to right as an example). Agents have an opinion, which is a position on this opinion line, and an interval ϵ demarcating the space around their position from which they take other opinions into account. Agents interact with each other in each step of the simulation. If the person they interact with is within the ϵ interval, they adjust their opinions closer; if the person is outside of ϵ then no interaction takes place. The Bounded Confidence model shows how opinion dynamics can lead to consensus building, radicalisation, or opinion plurality, depending on starting positions of agents and the size of the influence interval ϵ (see Figure 1.5).

With over 4000 citations, the Bounded Confidence model is the most cited and possibly most explored agent-based model to date. A similar model capturing opinion dynamics can be found in Deffuant et al. (2002), where instead of the bounded confidence, agents have a level of uncertainty in their own opinion. This model helps to understand extremism and radicalisation dynamics by assuming that those on the extremes are less uncertain of their opinion than those in the centre.

Instead of a line, agents can be placed on a grid and interact with their neighbours, gradually adopting political opinions that are influenced by those around them. This process mirrors real-world opinion dynamics, where individuals are swayed by the people in their social circles, leading to clustered regions of political leaning similar to how residential segregation patterns emerge in Schelling's model (see for example Wilensky, 1999b). It is important to stress here that while the final result of the segregation and the voting model look the same, the processes underlying the model are very different, with no agents changing themselves in the segregation model and no agents moving in the voting model.

The overall dynamics of social influence underlying the opinion dynamics models can also be adapted for other social influence mechanisms, such as the establishment of social norms (Elsenbroich and Gilbert, 2014, Ch12).

1.6 Empirical Agent-based Modelling

These classic models represent some basic underlying dynamics of human interactions, spatial, strategic, exchange, and assimilation. They are however rather abstract models, representing human behaviour in highly stylised ways. There has been plenty of criticism of these so called "toy models" as not applicable to real world situations (Moss, 2007). Increases in computational capacity, together with the availability of vast amounts and new kinds of data, has led to a focus on empirical agent-based modelling that integrates real-world data and evidence into different aspects of model development.

Models can use more empirically grounded theories to design the agents and their behaviours. Rather than starting with a stylised theory of human interaction, as underlying the four classic models discussed, 'common sense' or psychological theories are invoked to operationalise agent decision-making.

At the forefront of the common sense approach to behaviour operationalisation lies the Beliefs, Desires, and Intentions agent, or BDI agent (Bratman, 1984). This framework was adopted from early artificial intelligence research. Beliefs capture the agent's interaction with the environment, how the outside world informs the agent's decision making. Desires capture the idea that any action needs a motivation. Intentions operationalise the agent's capacity to form plans for actions. Extensions to this framework can be found in agent-based models that are influenced heavily by engineering considerations (such as Dignum, 2021).

More often, in particular in models in the social sciences, agent behaviours are informed by theories of human behaviours. Importantly, we need to distinguish between theories like game theory that are mathematical theories and theories derived inductively from data, like most psychological theories.

A commonly invoked psychological theory is the Theory of Planned Behaviour (Ajzen, 1991). With a similar structure of Intention leading to Behaviour, the Theory of Planned Behaviour disentangles the beliefs and desires part of the BDI agent into attitude, social norms, and perceived control. A key difference is explicit inclusion of a societal dimension into the agent's decision making by invoking norms. In contrast to the BDI agent, the Theory of Planned Behaviour has a strong empirical grounding (Ajzen, 2011).

A third way to elicit behaviours for an agent-based model is to collect new data. Asking people how they make particular decisions often makes agent-based models much more realistic than depending on a more generalised theory (even if the theory is empirically corroborated). For example, Gilbert et al. (2009) asked estate agents how they value houses that are coming on the market and implemented the answers directly into the agent-behaviours of the model.

There are several reviews of approaches to decision making in agent-based modelling (Balke and Gilbert, 2014; Wijermans et al., 2023) and many different meta-frameworks to help decide what theories to invoke (Schlüter et al., 2017) or what kinds of capacities agents need for a particular model (Verhagen et al., 2023). But it is not only the cognitive and social aspects of agent-decision making that are potentially important for agent-based modelling. Understanding the situational nature of decision making, how the physical and social environment might influence an agent's actions, is a still nascent field of research (Ncube et al., 2024).

While the behaviour of the agents is one of the most fundamental decisions in an agent-based model, many other aspects of an agent-based model can be more or less empirically implemented. You might want to have a highly naturalistic environment, based on a GIS map (see Chapter 11). Or you might have a highly realistic population, basing your agent-based model on a complete population dataset or a synthetic population (Chapuis et al., 2022)

Models can be heavily driven by quantitative data. These models use observed data, such as geographic information systems (GIS), demographic statistics, or real-time behavioural data to drive the model's social or environmental processes. These models can help researchers analyse simple human interactions in combination with environmental factors, making them useful for epidemics, transport or urban planning.

These kinds of models are often connected to geography, emphasising the importance of the agents' interactions with a specific environment and integrating data about the topic (Heppenstall et al., 2021). One major research area is traffic and pedestrian movement models that include real-time traffic data and historical accident reports. These might be used to help plan commuter traffic in cities or even evacuations during disasters (Malleson et al., 2022). A second major application area for this kind of quantitative, empirical agent-based model is crime and crime displacement (Birks et al., 2025). These models are again based on GIS, integrated with crime statistics, and demographic data. A third large application area is land-use and land-use change (Noszczyk, 2019). These models simulate how landowners make decisions about land-use in response to economic and environmental changes and can inform sustainable land management policies.

Behaviour, population characteristics, geographical features, patterns of movement, and activity measures provide different empirical foundations that may be more or less relevant for specific agent-based models. Unfortunately, a lot of data needed for modelling may not be readily available and primary data collection is often necessary. In fact, informing future data collection is one of the often cited uses of agent-based modelling (Edmonds, 2017; Barbrook-Johnson et al., 2017).

1.6.1 Modelling a Housing Market

An alternative to the geographical backdrop of empirical agent-based models is to implement an aspect of the economy realistically. Gilbert et al. (2009) develops a model of the housing market in England and Wales. Here the macroeconomy is modelled as a background system with highly realistic parameter settings (e.g. income distribution, interest rates and loan-to-value ratios etc). The agents are people who own houses or want to buy houses. Some people who own houses want to sell houses. There is one other set of agents called 'realtors' who are estate agents, valuing houses, effectively setting the initial sale price.

The behaviours of agents are implemented following realistic behaviours of, for example, putting a 10% markup on house prices at the point of putting them on the market or wanting to sell at a maximal price. There is no particular geography in this model, no map of the United Kingdom or a city in the UK. It is a model of the housing market in England and Wales because the agent behaviours are modelled on the behaviours of buyers and sellers in this country. The main peculiarities of housing market behaviours in the UK are:

1. People want to buy the most expensive house they can afford: The most expensive house they can afford is determined by both their wealth (which determines the deposit they can put down) and their income (the repayments they can afford). From those two the maximal mortgage they can get is calculated.

2. People move house when they can upgrade due to changed finance: In the UK people climb what is called the *housing ladder*. People enter the housing market by buying a relatively cheap property. As they pay off the mortgage on this property and have a higher income due to for example salary increases, they put this property on the market to upgrade to a better house.

The model replicates many features of the real housing market. It demonstrates the emergence of more expensive and cheaper areas (see Figure 1.6). It also shows many of the expected reactions of the housing market to macro-economic changes, such as interest rate or loan-to-value shocks.

The latter demonstrates a good use of the model. A sudden change to loan-to-value ratios leads to a severe drop in house prices, which, however, also recover relatively swiftly. The model lets us investigate why this is: The change to the loan-to-value ratio leads to first-time buyers not being able to enter the market as they do not have enough deposit. This in turn means that agents (owners) that want to upgrade do not have a buyer for their house. This lack of upgrading cascades through the system, leading to temporary stagnation in transactions, resulting in a price crash. Once the prices have crashed sufficiently for first-time buyers to enter the market again, the market recovers.

This housing market model is a good example of an empirical agent-based model but one where quantitative data is not the main driver. It shows that the need to stipulate the processes that drive change in the model does not mean you have to leave the realm of empiricism. The model has been further developed, broadening the agent-behaviours to include a private rental market (Gamal et al., 2024).

1.7 How to use this book

You are presumably reading this book because you have seen an agent-based model (or read about one) and decided that agent-based modelling is the most appropriate simulation methodology to help you investigate some research question. Or maybe one of your colleagues has suggested it? Unfortunately, there are few courses available, and many people building their first model are doing so with limited programming experience and without the assistance of experienced colleagues. If you are in this situation, this book is intended for you.

The core of the book is a tutorial that will gradually take you from a 'blank page' (what you see when you open NetLogo) to a fully developed model. It is essential that you actually do the tutorial and engage intellectually with the material. This book will not help you learn agent-based modelling or NetLogo if you simply read it or if you do the tutorial without reflection.

The book is structured like a course rather than a textbook or reference book. Like an in-person course, there is a certain amount of foreshadowing of future material and reviewing material already presented to help you connect what you are doing in the tutorial to the broader theory of agent-based modelling.

Key points

Boxes like this contain definitions, concept summaries, programming tips, and other important points. They do not, however, describe NetLogo language keywords, which are displayed in a different type of box.

While the tutorial model was developed to use many of the most important parts of the NetLogo programming language, it does not cover all of them. In particular, the agents in the tutorial

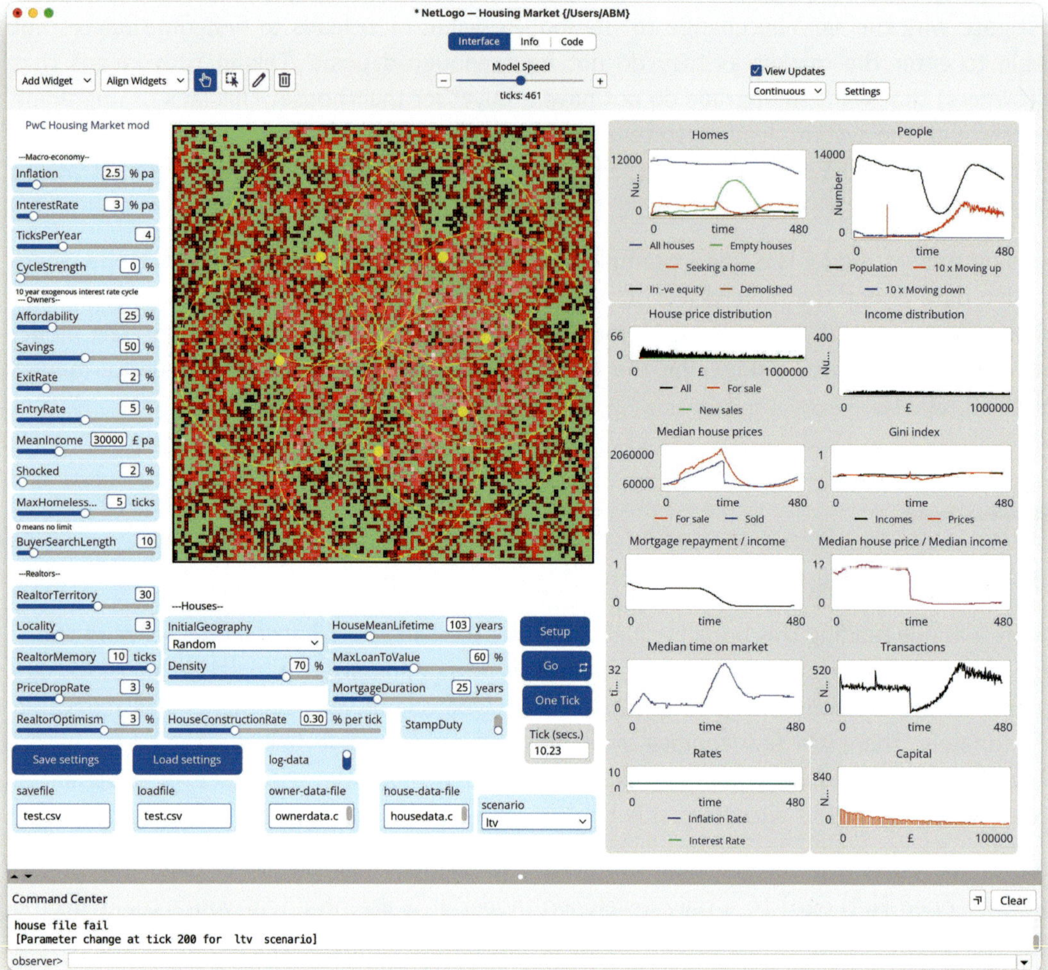

FIGURE 1.6: Model of the UK Housing Market: This is an interface generated from the model of the UK housing market described in Gilbert et al. (2009). The model was run for 100 steps with a loan-to-value ratio of 100%. The loan-to-value ratio was then reduced to 60%. The view in the middle shows the emergence of more and less expensive areas, with light pink being more expensive. The graphs on the right hand side show the house price crash after the loan-to-value ratio change and the uptick (top) as well as the market stagnation, i.e. the time houses are on the market (bottom). The blue sliders on the left show a range of input parameters into the model, demonstrating the level of complexity.

model do not move, so mobile agents are dealt with separately (see Chapter 11). Other chapters present some more advanced NetLogo topics that are useful for certain model needs. These topics either use models from the library included with the NetLogo software, or are presented as standalone examples: small pieces of code that can be run within NetLogo but are not part of any model.

As a result, there are many pathways to use this book and the choice of pathway depends on the purpose of your intended model. Some questions to help your choice are presented after the tutorial model (see Chapter 10), at which point the questions will hopefully make more sense to you.

1.8 Summary

We have discussed societies as complex systems, complex systems and their properties, modelling and simulation, and agent-based modelling as a good method to model complex social systems. We furthermore introduced four classic models that exemplify fundamental processes underlying social phenomena: movement, strategic interaction, exchange, and social influence or assimilation.

Agent-based models are particularly useful for studying systems with heterogeneity, interactions, non-linearity, and emergence — all core aspects of social complex systems. They allow researchers to move beyond the limitations of static descriptions, capturing the dynamic and evolving nature of social phenomena. Agent-based modelling enables us to investigate how micro-level interactions can lead to macro-level changes, providing insights into everything from urban segregation to market dynamics to political polarisation.

In summary, understanding societies means understanding complex systems. To do this, we must account for the heterogeneity of agents, their interactions, and the dynamic processes that produce macro-social phenomena. Agent-based modelling offers a powerful way to investigate these complexities, revealing patterns and dynamics that traditional methods may overlook. While complexity theory in the social sciences extends beyond agent-based modelling, the models provide a critical lens through which we can explore the forces that shape our world.

Questions

1. What is a complex system?
2. What is a model? And what is a simulation?
3. Why is agent-based modelling a particularly good simulation method for complex systems?
4. Think of a social phenomenon that is driven by interactions. What characteristics of complex systems does it display?
5. List three attributes of complex systems that agent-based modelling is good at understanding.

2

Building an Agent-based Model

This chapter covers the core concepts of agent-based modelling, including agents, environments, and interactions. It explores key challenges in model design, such as system boundaries, abstraction, idealisation, and operationalisation, and highlights the generative purpose of ABM in social science, moving beyond traditional views of causality. The chapter positions ABM as a methodology for understanding complex social systems, capable of supporting diverse models that, despite varying in detail and purpose, share a focus on replicating essential processes to explain real-world patterns.

2.1 What is an Agent-based Model?

The previous chapter introduced societies as complex systems, simulations as a method for modelling them, and agent-based modelling as a particularly good modelling method as it most closely mimics the interacting parts and processes underlying complex phenomena. We have also seen some classic agent-based models, which exemplify basic social processes, like segregation, strategic interaction, exchange, and social influence. While we defined the terms *model* and *simulation*, we have not yet defined an *agent-based* model properly.

Agent-based Model

An agent-based model is a computational simulation model of autonomous agents interacting with each other and an environment.

In this chapter we will get more specific about agent-based modelling, focus on its peculiarities and what these models allow us to represent. We will also look at the research process, the decisions to be made to build an agent-based model, and the purpose of agent-based modelling.

Let us first go through the ingredients of an agent-based model and how they relate to the relevant features, relationships, and processes we need to represent.

2.1.1 Agents

The first ingredients to represent in an agent-based model are the agents. Agents are the fundamental components of the model. Agents in agent-based models are often individuals, but they can also be composites, such as households or organisations (Hailegiorgis et al., 2018; Gilbert et al., 2014), concrete things like houses (Gamal et al., 2024), or abstract concepts such as social practices (Narasimhan et al., 2017).

Agent

An agent is an autonomous entity with a set of attributes and behaviours.

Agents have attributes, e.g. age, sex, and income; blue or red; politically on the left or right; being a cooperator or defector. This possibility to model a range of attributes allows agent-based models to represent heterogeneity across a population (similar to microsimulation). Agents also have autonomy, they make decisions about how to act. An agent in an iterated prisoner's dilemma might have an initial preference to cooperate or defect (the agent's attribute), but during the simulation it makes an autonomous decision to cooperate or defect in each interaction. This decision might be made at random or be informed or determined by an agent's current situation or past history or a combination of both. Given that preferences can change depending on an agent's experiences and contexts, levels of heterogeneity across the population can also change over the run of the simulation. Think for example of an opinion dynamics model leading to consensus over time (see Figure 1.5c).

Agents come at different levels of cognitive complexity (Balke and Gilbert, 2014). The level needed depends on the system modelled. For an agent in a simple segregation model (Section 1.5.1), it is enough that the agent can perceive its surrounding patches and calculate the ratio between the different types of neighbours. For more complicated segregation dynamics, for example involving school choices (Sage and Flache, 2021), agents need a more complicated utility function, taking into account internal, i.e. its own preferences, and external attributes, e.g. school profile, travel distance etc. Make sure your agent's cognitive capacities are commensurate with the behaviours you are modelling. There is no point in using a highly cognitive agent architecture to implement simple decisions.

The key abilities that enable agents to interact with their environment and other agents include autonomy, communication/exchange, mobility, and the capacity for learning and adaptation.

1. Autonomy: Agents are self-contained and make decisions independently based on their internal state, rules, and interactions with their environment. This allows them to initiate actions and achieve their goals without direct external control.
2. Communication/Exchange: Agents can communicate with each other and their environment, exchanging information or resources and coordinating actions. This can involve for example querying other agents or searching for specific attributes.
3. Mobility: Agents can move around in the simulated space. This mobility, coupled with communication and perception of their situation, enables a wide range of interactions and behaviours.
4. Learning and Adaptation: Agents can be designed to learn and adapt their behaviours based on past experiences or feedback. This can involve altering their internal states, modifying rules, or even evolving at the population level (using some fitness function and selection mechanism).

2.1.2 Environment

While the agents are the main source of heterogeneity in the model, heterogeneity can also be encoded into the environment. We have already seen in the Sugarscape model that resources can be distributed unequally across an environment (Section 1.5.3). The combined heterogeneity of agent attributes and environments allows agent-based models to represent context and variation in contexts in a natural way.

The environment can represent many different things in an agent-based model. The models discussed in Chapter 1 had the environment as homes for agents in the segregation model, resources (sugar and spice) stored in the environment in the Sugarscape model, an opinion space in the Bounded Confidence model, and a simple random physical space in which agents would meet to play a Prisoner's Dilemma game by proximity. An environment can also be completely a-spatial, consisting of a network of links between agents, for example representing a friendship network between people or trading relationships between firms (see Chapter 8).

Sometimes there is no intrinsic need for an environment. For example, the Bounded Confidence model works equally well without the representation of political opinion on an opinion line; the line is a visualisation tool. Other models rely heavily on the possibility of representing spatial features, for example urban planning and evacuation models (Crooks, 2016).

As for agents, when thinking about the environment, think about what you *need* to represent for the model to work, and what is superfluous. Does your model need a realistic GIS map underlying the simulation? Or is it enough that you have some geographic features of the terrain, like mountains and rivers; or can you even leave that out and have a completely stylised environment? Just "having data" is never a good enough reason to put something into the model.

2.1.3 Processes

A process is how a system steps through time. As discussed briefly in Section 1.4.4, there are two ways of representing processes. Hartmann (1996), where we took our definition of simulation from, defines *process* as a description of the "temporal sequence of states of a system". This can be seen as an external view of the process of system change. It is sufficient to understand *that* the systems changes and (possibly) *by how much*, to reproduce the measurable changes of system variables (e.g. time series methods such as regression models).

Another way to think about process is internal to the system, representing *how and why* the system changes rather than just *that* and *by how much*. This means to understand some of the causal mechanisms at work that bring about new system states.

The simulation methods discussed in Section 1.4 all look a little bit under the hood of this system-level view of process. Discrete event simulation explicitly models component changes as events that bring about system change. In system dynamics, stocks and flows represent processes at the aggregate level. Microsimulation uses transition probabilities to model state changes for individuals, thus helping understand how micro-state-changes lead to overall system change.

Agent-based modelling represents the mechanisms underlying systems changes; instead of just modelling individual state changes by transition probabilities, agent-based models represent *why* changes to individuals happen. For this, agent-based modelling adopts the perspective of the agents within the system. The model comprises many agents and, at each step of the simulation, each agent executes a set of actions (which could be doing nothing!), depending on its personal characteristics, current and past situation, and possibly anticipated future events. Importantly, each agent makes this decision of action *autonomously*; there is no central controller making the decision for them.

This *agency* is the first and foremost driver of an agent-based simulation and constitutes the actions (and interactions) of the parts of the complex system. Agent-based models can also incorporate other dynamics, for example, have an underlying microsimulation to model demographic change over time or have a system dynamics model drive changes in the environment.

Agent-centric perspective

I, the agent, have certain characteristics and beliefs of my own as well as information about the world around me, and therefore will decide on some action.

The implementation of this perspective in code is best understood with specific code examples, so it will be revisited throughout the tutorial. At this stage, it is sufficient to recognise that agent-based models represent the target system from an agent-centric perspective.

This agent-centric perspective is also the hardest part of designing an agent-based model, as for a lot of actions and choices we do not exactly know how individuals make them. Do you always know exactly why you ate that burger, went for a run, or bought that pair of jeans? And if we do not even know it for ourselves, knowing exactly why other people make particular choices is utopian. So, while we are trying to represent the choices of real people, our computational agents will be simplified approximations, or (abstracted and idealised) representations, of real people. These representations will take into account information about the world to stipulate what makes an individual more or less likely to perform an action. We know we are more likely to go for a run when it is not raining or eat a burger when a fast food restaurant is in close proximity.

2.1.4 Interactions

We have already seen in Section 2.1.3 that agent-based models are driven by individual agents executing actions. But actions have consequences. Me moving to a new neighbourhood might make you unhappy and move away in turn. Me eating all the sugar leaves none for you. Me changing my political opinion might move us too far apart for you to influence me. Me defecting in a prisoner's dilemma means we cannot get the high score together, and might make you defect in the next round. Actions in an agent-based model are *interdependent*, with other agents' actions (you defecting in a prisoner's dilemma) and with the environment (e.g. us depleting resources).

It is these interdependencies that drive the overarching processes of the model, the macro-processes. It is also these interdependencies that make the agent-based model a model of a complex system (as described in Chapter 1). It is the continuous change — reassessment — decision cycle, played out by each agent at every step of the simulation, that brings about the necessity of a computer simulation, as it is too difficult to compute this mentally.

Let us go back to the segregation model. Schelling initially executed this model on a checkerboard with different coloured coins (Schelling, 1971). Every move of one unhappy agent leads to a change in the environment of up to 16 other agents (its eight new neighbours and its eight previous neighbours). On a 8 x 8 board with maximally 63 "agents" (one field must stay free to allow for movement), the simulation is just about tractable by hand. As soon as these numbers increase, a computer is a welcome thinking aid.[1]

[1]Although initially Schelling was vehemently opposed to using a computer to understand the model (see Hegselmann, 2012).

While the emergence of segregation can be seen on this small grid, a larger grid is less sensitive to initial conditions (e.g. density) and better demonstrates the model's tipping point of wanting just more than a quarter of neighbours to be similar to oneself. And while more often than not the general phenomenon explored in an agent-based model does not need millions of agents to be represented, it needs an adequate population to interact with each other.

Computer simulation can help us expose consequences of interactions that are not immediately obvious (unforeseen consequences). For example, in the housing model in Section 1.6 the housing market crash following the reduction of the loan-to-value ratio was a surprising consequence. The model helped us to understand that it was the lack of first-time buyers, who could not enter the market as they did not have enough deposit, that led to properties further along the property ladder remaining on the market for longer, reducing their prices.

2.1.5 So, what is an Agent-based model?

Putting these elements together constructs a more detailed definition of an agent-based model. It is a computational model of a process from the perspective of the agents. The relevant features and relationships that are to be included in the model are those that contribute to the decisions and behaviours that the agent must take in order to represent their equivalent real-world entities. Furthermore, characteristics that have no effect on decisions and behaviours do not need to be included in the model.

Agent-based Models (Detailed)

An agent-based model is a computer program that represents individual entities taking actions in accordance with their own characteristics, resources, beliefs, and perception of their social and physical environment.

While agent-based models are typically run on a computer, that part of the definition is descriptive rather than prescriptive. This is still a minimal definition of agent-based models. Other descriptions of agent-based modelling may also include non-definitional aspects of the method, such as experimentation with different scenarios (Gilbert, 2008), or advanced agent features such as the capacity to adapt (Macal and North, 2010).

This slightly more detailed definition of an agent-based model suggests a list of things that one needs to think about when building such a model. What are the important characteristics, resources, beliefs, and perceptions of an agent? And what is the social and physical environment? How do all of these aspects interdepend? And how do we best represent these in a model?

Conceptualising an agent-based model makes us ask ontological, epistemological, and methodological questions. We have to ask ourselves what the entities and processes are that we are representing in our model. By this decision we are making existence statements about the real world, asserting correspondence between the model and the target system. We are also making epistemological assertions, that these entities and processes are the ones relevant to understanding the particular complex sub-systems we are interested in. We also contend that we can know the relevant aspects of the system, that we have methods to find out about them.

This is one of the crux points of agent-based modelling: We will have aspects of the model we do not know — either by ignorance (we have not collected the data yet) or in principle (the kind of data or evidence cannot be collected (e.g. because it is not measurable)). It is important to keep track of these decisions on representation, which aspects have direct counterparts in the real world, which aspects might have to be stipulated for the model. For example, to run the dynamics of the model we might need a *threshold*, a value above which agents change their behaviour. While in the model this threshold will be represented by a number, say 0.75, there is no corresponding 0.75 actual threshold in the real world. The threshold expresses that it takes 'quite a lot' for people to change their behaviour.

An agent-based model is a representation of some aspects, a subsystem, of the complex system that is the real world. This representation is carefully crafted, more akin to a painting than a statistical model. While containing any amount of quantitative data, the method of agent-based modelling is more qualitative, a sense-making endeavour rather than a summary of collected data. We return to this theme in Section 2.5.1.

2.2 Abstraction and Idealisation

A model is a representation of the target system. A social science agent-based model is a representation of a social system. It is not a replica, a smaller version of the original. It does not have real people in it, and does not take place in a physical environment. It is a representation, a depiction of the aspects of the social world relevant for the research question. That means it leaves things out, simplifies the system to its relevant components. We have now stumbled over some concepts again and again, concepts like representation, operationalisation, and abstraction. These are all important activities in modelling.

A *representation* is a depiction of something. It is not the thing itself but an object recognisably standing in for the real thing. René Magritte produced some famous highly naturalistic paintings of a pipe and an apple; written on the paintings are the sentences "Ceci n'est pas une pipe" and "Ceci n'est pas une pomme". Magritte wanted to bring home the message that the paintings were *representations* of those objects and not the objects themselves. The paintings are two-dimensional, cannot be smoked or eaten, do not deteriorate and ferment. We can learn what the objects look like from the paintings, but that's it. When thinking about models we have to remind ourselves that the models are representations of the real world, leaving out dimension, functions, and contexts.

While an agent-based model is a very naturalistic representation of the real social world, just like the realistic paintings of Magritte, it leaves out many of the features of the real object, the complex system it is representing. We can distinguish two different kinds of *leaving out* features in modelling; abstracting and idealising.

Abstraction

Abstraction is the process of stripping away (irrelevant) representational features.

A model has to represent the *relevant* features and relationships, those aspects of the real world that have causal impact, that we think make a difference within the target system. That means that aspects that are irrelevant can be left out. For example, we rarely give agents hair colours or star signs; they do not make a recognisable difference to people's behaviours. But, if we were to devise a model in which partner choices of agents were driven by star sign compatibility, we would not be able to abstract it away. Note that the decision of what is relevant is in relation to that specific model; for another model previously abstracted features might be highly relevant. For example, a model of the labour market focused on economic growth might not have gender as an agent variable. A model of the gender pay gap in the labour market cannot abstract gender away.

It is important to see abstraction as driven by the modelling purpose, not the availability of data. This means that we neither strip away aspects just because we do not have data, nor include aspects just because we have data.

Once we have decided which features can be abstracted and which need to be included, we have to think about how to include these features, their *operationalisation*. For some features we might need to make sure we have exact data, but more often than not it will be sufficient to idealise: for example, to have an income distribution curve of the right shape rather than the exact income distribution of a country.

Idealisation

Idealisation is the process of smoothing irregularities of representational features.

The possibility of idealisation is important for agent-based modelling in particular, as there will be features in a model for which there are no data available. For those features, having an idea of an idealised version of the features allows us to still operationalise and include those features. Think for example of the operationalisation of an agent's political opinion as a number on a continuum between 0 and 1 and visually as a position on a line.

When designing your agent-based model, make sure you consider what relevant features the model needs to represent, and strip away those that are not relevant. Operationalise those that are included by either real-world data or idealisations.

2.3 Decisions, decisions, decisions

All research starts with a question. And it is the questions you are asking that determine the choice of method. There is no point in doing a set of interviews or focus groups, at least not on their own, if you want to know the distribution of self-rated physical health across the population of the United Kingdom. Similarly, you would not do a survey on a representative sample of households across the whole of the UK if you want to know about the effects of flooding on mental health and wellbeing, as only a small percentage of people has ever been flooded. You would either do a targeted survey or interviews and focus groups, depending on the kinds of questions you are trying to answer.

Like all methods, agent-based modelling is a method that is excellent for investigating certain questions and entirely inappropriate for others. A simple way to exclude agent-based modelling from your potential methods is that if you find it difficult to define agents, environment, or interactions, then agent-based modelling is not the method you should choose. But even if you have individual-level data to define your agents and a good idea of an environment, agent-based modelling might not be the right method for your project. If you do not know anything about the behaviours of your agents, the mechanisms of interaction, you will not be able to build a good model. Having a dataset with individual attributes is not enough to build an agent-based model. The crux is the agent-centric processes that you need to specify to implement an agent-based model. Your model must represent *how* behaviours of agents and their interactions lead to change within the system. Again, make sure you are not led by data availability alone.

2.3.1 What to do? Where to do it? And what else is there?

There are two ways of approaching behaviours in agent-based modelling. One is to stipulate a behaviour or set of behaviours as an explanation for a social phenomenon. There is no attempt at realism, but rather a stipulation of a scenario that looks at the potential of how simple rules of behaviour can lead to complex aggregate outcomes. For example in the Prisoner's Dilemma simulation, the investigation is into whether (and under what conditions) an iterated Prisoner's Dilemma played across a diverse population brings about cooperation at the macro-level, although the individual's single shot best strategy is to defect. The Prisoner's Dilemma here stands in for any strategic interaction with asymmetric payoffs. For the Schelling model, simply the possibility of people moving if they do not like their neighbourhood can explain high levels of segregation despite relatively low similarity preferences. Of course, in the real world we are not playing daily Prisoner's Dilemmas and people cannot just move house willy-nilly.

What these models tell us is not that it is these exact processes that explain societal cooperation or racial segregation, but that the cumulative outcomes of these interactions between agents lead to surprising outcomes related to the original research question. The upshot of the Segregation model is not that there is no racism and that segregation is just a result of mild preferences. Instead, it is that segregation is hard to completely eradicate even without the presence of racism. These models are stylised ways of investigating the interactions and interdependencies without much focus on realistically representing behaviours and choices.

Often called 'toy models', these stylised agent-based models have become heavily criticised for reductionism, misrepresentation, and oversimplification. Edmonds and Moss (2004) argue against the focus on simple models, highlighting the need to be descriptive (summarised in the "Keep It Descriptive, Stupid" (KIDS) paradigm). In a similar way to KIDS, "Enhancing the Realism Of Simulation" (EROS) espouses a realism, rather than simplicity, paradigm (Jager, 2017).

This focus on realism leads to a second way to implement behaviours, trying to replicate as closely as possible people's behaviours in the real world. This can be done in many ways; from using common-sense approaches to decision-making, invoking psychological theories, to developing behaviour rules from focused qualitative data collection (see Wijermans et al., 2023, for an overview of approaches). The choices of theory detail and agent architecture (see Section 2.1.1) are interdependent, as you need more cognitive capacity for agents that socialise, adapt, and learn.

Even if you know about the behaviours that you want to implement, you might not know about the decision-making process for agents, i.e. why they behave as they do. If you are interested in a model of physical activity, the agent's decision to go running will be determined by its utility, habit, and fitness level in some ways, but in the real world myriad aspects will play into the decision to go running on any particular day, including how the person slept the night before, the weather, or whether something unexpected happened to reduce time available for exercise. We cannot include all these aspects into the model so must find the right level of abstraction and idealisation (see Section 2.2) to make the model work.

Like with everything in agent-based modelling, we can think about the choice of representing human behaviours as a continuum of abstraction. We always replicate the behaviours as realistically as necessary for the model, not as realistically as possible. There is no point choosing a highly cognitive agent when implementing an epidemiological model with simple behaviour rules. The choice of how closely human behaviour needs to be represented is driven by the need of the model, not by the provision of data from the world.

While opting for more complex behaviours might enhance construct validity of the model, it might also make output validation more difficult (see Section 14.4 for more details). A model is there to better understand what is going on; agent-based models in particular are models that want to open the 'black box' (Dilaver and Gilbert, 2023). In a more complicated model it is harder to fully understand the interactions and processes and how they lead to the outcomes of interest. So the mantra should not be "Keep it Simple, Stupid (KISS)", but "Everything should be as simple as possible, but no simpler" (a quote attributed to Einstein).

For the agents and behaviours we have to make decisions about how complex and realistic they should be. This question is also pertinent for the environment. First of all, we have seen that the environment can be something completely different from a physical environment, e.g. an opinion space. For physical environments the question of how realistically they have to be represented is more pertinent. Section 11.6 introduces how to integrate real-world GIS data into your agent-based model. While this is a great feature of agent-based modelling, and will be essential if you want to build something like an evacuation model of a specific coastal region in a flood, most agent-based models will not need that level of realism in the environment.

Just as we had with the agent specification, only implement what is necessary for the model. If the exact geography does not make a difference in your model, you do not have to implement the geography. This does not mean that geography does not make a difference in the real world. But in order to model the difference geography makes in the real world, it is not sufficient to upload GIS data into your model, you have to specify *how* the geography will make a difference within the model you are building. For example, in the housing market model described in Gamal et al. (2024), agents buy the most expensive house they can afford, given their capital and income. The environment is completely stylised, initialised with geographically random house prices. Over the simulation run more expensive and cheaper areas emerge. Of course this model could be run on a GIS map of any city, region, or country where data is available, but it would not make any difference to the model as agents do not incorporate questions around transport links, amenities, or nice housing stock into their decision-making, so the real geography is superfluous. The model can be adapted, of course, to use for example GIS data to initialise more expensive and cheaper areas explicitly in accordance with the target system rather than having these areas emerge.

You might want to represent other entities in your model: houses, schools, food, money, or even abstract things like advertising campaigns, norms, and social practices. Some of these other entities could be represented through the environment; for example houses could simply be locations on the grid, thinking of a house through its unique address identifier. Or they could be represented as agents, where the house is placed in a location on the grid but can be demolished and a new house built in the same place. Abstract entities can also be implemented in different ways, for example as behaviour constraints to agents, or as in Narasimhan et al. (2017), where social practices are themselves agents.

This freedom of conceptualisation and operationalisation of agent-based models is simultaneously a strength and weakness. The construction of an agent-based model demands creativity and ingenuity on the part of the modeller, who has to think hard about how to best represent the relevant features and relationships of the real world. Good choices are not always clear. That does not mean they are unscientific; agent-based models help you to tell empirically based stories about phenomena in the real world (see Badham et al., 2021).

2.3.2 Decisions have Consequences

Your decision about what to put into the model is a decision about the boundaries of the system you are modelling and not necessarily a full ontological statement. You can acknowledge that many other things and features exist in the real world (ontological commitment) but leave them out of the model in the process of abstraction. This abstraction is an assertion that these features are not (causally) relevant for the particular features, relationships, and processes you are replicating.

Given the chance that we might abstract away features that are in fact relevant to the research question, maybe we should not abstract at all. The reason for abstraction, at the peril of getting it wrong, is that the more complex your model, the more you lose the ability to completely know and understand your model. A map the size of the forest will not help you to find your way. In particular for agent-based modelling, tractability is essential due to its purpose of opening the "black box" and helping us understand how processes at different levels (micro-meso-macro) lead to observed phenomena (Dilaver and Gilbert, 2023).

Increasing complexity and realism have consequences for modelling practice and application. Increased complexity means you will need more data to calibrate your model. Every attribute or feature you add needs to be considered as having an operationalisation that keeps it as representative of the real world as necessary for the model. Every parameter you add to your model is also an increase in the size of the parameter space to be explored for analysing your model.

Increased realism means you need more detailed validation, checking whether the model corresponds to the real world. A highly abstract model, like the ones we have seen in Section 1.5, only needs to show they replicate maybe just one single pattern, like the bifurcation of 'opinions' into extremes in opinion dynamics. Less abstract models, also often called mid-range models, need to show that they replicate combinations of patterns, for example market crashes and recovery over time given particular macroeconomic settings for a housing market model. Highly realistic, or facsimile models, might well need to show that they replicate patterns together with magnitudes found in the real world, for example how long the evacuation of a certain number of people will take from a building.

If you want to illustrate an underlying dynamic, it might be enough to build a highly abstract model replicating that one dynamic, whereas if the goal is policy advice the model will need much higher levels of realism. The protocol developed in Grimm et al. (2006) can be used to structure thinking about the questions in Section 2.3 as can the framework developed in Dilaver and Gilbert (2023). In this book we are using a simplified table to help thinking about modelling decisions; see Table 2.1.

TABLE 2.1: Table of broad design questions

Aspect	Question
Process	What actions or repeated decisions?
	What state changes without action?
	What does theory say about these processes?
Agents	What entities take actions?
	What personal characteristics affect action?
	Any other entities need to be included?
	What characteristics of those entities are relevant?
	What characteristics change over time automatically?
Spatial	Does location change the way in which the process operates?
	What location specific resources (including information) affect action?
	How do resources change over time automatically?
Social	Do other entities change the actions of individuals?
	How are agents aware of each other?
AA interaction	What is the interaction between agents?
AE interaction	How do the agents affect the environment?
EA interaction	How does the spatial environment influence actions?
EE interaction	How do the characteristics of a location impact on other locations?

The decision of what kind of model to build is somewhat dependent on your research question but might depend even more on the overall research the model sits within. Note that agent-based models do not need to be the end product of a research project. They can be a part of a project, contributing to evidence and data synthesis or used to test certain hypotheses. A case in point is Castellani et al. (2015), in which a simple agent-based model is used to test the hypothesis that lacking health improvements in deprived areas, despite high levels of investment, are a result of better-off residents leaving the area as soon as they can ('suburban flight').

2.4 The Modelling Process

Any research project has a set of stages; and projects involving agent-based modelling are no different. In the last section we went through lots of decisions you have to make conceptualising, designing, and implementing your model. But the model building itself is only one part of a wider modelling project. In this section we will take a step back and look at the research process for modelling and the overall research an agent-based model sits within.

Any research starts with a research topic, an area of the real world that one wants to understand better; the target system. In order to build a model, we need to produce a specification of the target system, a description of those relevant features and relationships. We then need to operationalise this description into a static model, or, for a simulation, a dynamic replication of a process.

We have discussed already that a model is a partial representation of the target system, the modeller making decisions about what is included and excluded from the representation. How much is included or excluded depends not on the topic area but on the kinds of questions one wants to answer about the target system.

Let us think about that model of the housing market discussed in Section 1.6 (or see Gilbert et al., 2009). We can start with some very general questions, like "How come some areas are more expensive than others?", or "Why do house prices go up steadily?", or "Why do we have house price bubbles?". We can ask some general questions about something specific, like, "What is the effect of certain macro-economic factors (inflation, unemployment, wages) on the housing market?". Or we can ask very specific questions about a specific housing market, like "What is the effect of certain Government policies on the UK Housing Market?"; policies like a "stamp duty holiday", 100% loan-to-value mortgages, or low interest rates. The choice of questions we will want to ask will determine the level of complexity and realism (see Section 2.3.2) of the model we need to build.

Once you have settled on the kinds of questions you want to ask, you can start thinking about your model: who are the agents?, what is the environment?, what do my agents do (how do people buy and sell houses)? You will also need to gather additional information about the real world, e.g. historic inflation rates or the income distribution of the population you are modelling.

Once you have listed and described your target system and gathered empirical information about it, you can start thinking how to operationalise this information for the model. This operationalisation has two aspects:

1. You need to specify how different aspects of the system interrelate. For example, the rate of inflation is important as it has an effect on disposable household income, which in turn has an effect on household savings and mortgage affordability.
2. You need to construct ways to quantify and implement a concept from the target system into a model. For example, estate agents have an area in which they operate, a particular part of town for example. In an agent-based model you might give an estate agent a location in your environment and demarcate a circle around them specifying their area of operation. This is where you have to do the abstractions and idealisations we introduced in Section 2.2.

Once you have made all those decisions, resulting in a verbal (or diagrammatic, e.g. UML) presentation of your model design, you can start to implement the model in code. In the process of coding you will inevitably stumble over aspects of your model that are underspecified and you will have to rethink your model, go back to gather data from the real world. This is completely normal, so do not despair. Agent-based modelling is a highly iterative process.

Once your model is implemented, it is now time to get to know your model. You might think that having built it you know exactly what is happening; this is not necessarily so. Experimenting with your model will help you with some general sense checking, for example checking that it still answers the original research question, as well as some more detailed checks, e.g. checking which parameters make a big difference. Just 'playing around' with the model, testing some extreme configurations, changing some parameters mid-simulation (e.g. testing the effect of an interest rate hike after 100 steps in the housing market model), and trying to find configurations where the model stabilises will give you a good overall understanding of the model. We will come back more systematically to experimentation in Chapter 13.

There are two more formal processes the model has to undergo: verification and validation. *Verification* is essentially checking that the model does what you intended it to do. This might sound trivial, but it is not. There are some obvious bugs that you might find quite quickly, resulting in syntax errors of the system or showing up as strange distributions in output histograms. However, sometimes bugs hide well within the code. In a model of extortion racketeering presented in Elsenbroich (2017), we initially found a very surprising outcome that extortion reduced swiftly at low levels of resistance. With a closer look at the code it became clear that this was a bug. Instead of "each extorter punishing an entrepreneur", all extorters punished the same entrepreneur. So when your model shows something interesting, your first call is to look into the model and find out the reason for this interesting result. Note, you are not looking for the explanation in the real world for this result, but you are analysing which *aspects of the model* lead to the result.

Once you have established that your model is correct, you need to check whether you have built the right model. This is the process of *validation*, of comparing the model to the target system it is supposed to replicate. We will discuss validation in more detail in Chapter 14.

Of course, having the model is not the be-all and end-all of the research. Your model design, implementation, experimentation, verification, and validation need to be used to actually answer the research questions you started out with. After all, the model was there to help us understand the target system, an aspect of the real world. In a research process involving an agent-based model, this 'use' typically involves a large number of simulation experiments (see Chapter 13). Being clear and honest about what the model helps us to understand about the real world is really important (and we will come back to this throughout the book). Always remember you are representing a sub-system of the real world, with abstractions and idealisations. And these limitations need to be communicated clearly when describing your model (see Section 14.6 for more detail).

2.5 Why Model?

We discussed in Chapter 1 that complex systems are difficult to grapple with and that the traditional way of carving them up into controllable units will not help us understand them. We advocated modelling, in particular agent-based modelling, as a method for researching complex systems. But what will the models actually allow us to do? There are two answers to this question: one more programmatic (Section 2.5.1), the other more pragmatic (Section 2.5.2).

2.5.1 Generative Social Science

Epstein (2006) introduces agent-based modelling as a computational approach to understanding social phenomena. The book establishes a name for the scientific programme of agent-based modelling: *Generative Social Science*. The programme can be summed up as explaining social phenomena by generating them *in silico*. It emphasises mechanistic, process-based modelling over purely statistical correlations, making agent-based modelling a powerful tool for studying social complexity. The core idea of generative social science is that to explain a social pattern, one must generate it from the bottom up using simple agent rules, rather than relying on traditional statistical or equation-based models.

Generative social science is a completely new paradigm. The persisting scientific conception of causality follows Hume's causal scepticism. According to Hume (1894), as we cannot experience causality directly with our senses we cannot make empirical causal claims. All causal claims are thus metaphysical statements. All we can claim empirically is that we observe a *constant conjunction* or a *regular association* of events. The strong empiricist position presented in Hume has influenced our scientific understanding of causality for centuries, focusing scientific efforts entirely on shoring up *constant conjunctions*. We have become very careful, asserting that all we can find are correlations, not causation.

Acknowledging complexity is a challenge to this conception of causality. In complex systems our traditional methods of eliciting causality, i.e. statistical regularity and experimentation, falter. One problem with this regularity-based approach to causality for understanding complex systems is that strict regularities are hard to find due to context dependency, equifinality, and multifinality in complex systems, meaning that generalisations will have many exceptions. The second problem is that emergence furthermore entails the impossibility of carving up complex systems into neat, controllable subsystems, which does not allow for the control and isolation that experimental methods demand. *Ceteris paribus* is an obsolete concept in complex systems as "all things being equal" is often not possible.

Generative social science deviates from this conception of causality. The paradigm states that if we generate the patterns we find in the real world in computer models, we have demonstrated a (potential) causal mechanism underlying the phenomenon, and can claim more than mere association. Generative social science can explain *why* things happen, it can tell the causal stories underlying observed correlative patterns. This could be a relatively simple explanation like segregation resulting from successive moves of agents away from neighbourhoods they do not like (Schelling, 1971), or a more complex explanation like a housing crash resulting from a change to loan-to-value ratios due to the inability of first-time buyers to buy, stagnating the "housing ladder" (Gamal et al., 2024).

So while there is a great variety of agent-based models, a point made repeatedly, the generative social science approach is a unifying feature of all of them. And it is this generative approach that needs to be kept in mind when building an agent-based model. Despite Hume's argument that we can only ever assert correlations in the real world, agent-based models are a tool to generate possible explanations as to how these correlations might have come about.

2.5.2 The purpose of agent-based modelling

The programme of generative social science sets out a purpose for agent-based modelling as generating social phenomena *in silico* and thus explaining the otherwise unobservable causal connections. The programme was summed up earlier in Epstein and Axtell (1996) as the change from 'Can you explain it?' to 'Can you grow it?'. Despite this, the question of the purpose of agent-based modelling is a live debate within the discipline (Elsenbroich and Polhill, 2023).

The reason for the persistence of the debate is one crucial potential purpose of modelling: that of *prediction* or *forecasting*. Despite their differences, the terms prediction and forecasting are often used interchangeably. Prediction is a general statement about a future event, often based on patterns or intuition. Predictions are often linked to particular events, such as elections or sporting competitions (Silver, 2012). Forecasting is broader, using data and models in a structured process to estimate future trends with probabilities in, for example, economics or the weather.

In traditional philosophy of science, explanation and prediction are often viewed as interconnected, a concept known as the "symmetry thesis" developed by Hempel (2014). It argues that both explanation and prediction share the same logical structure: if initial conditions and a general law hold, then a phenomenon can be both explained and predicted. Applying the symmetry thesis, the generative social science paradigm thus transforms into "if we can grow it, we can explain it, we can predict it". While this may apply to simple models and rather hand wavy predictions, like "we can predict high levels of segregation", the rise of empirical, data-driven agent-based modelling has shifted the focus to complex socio-ecological systems. As agent-based models become more relevant in applied social science and policy, explanation takes a backseat, making the role of prediction increasingly critical. The key question remains: Should this transitivity still hold for empirical, data-driven ABMs?

Epstein (2008) tackles the question about prediction by asserting that anyone who makes a projection about a social phenomenon, a pandemic, a war, or anything else, is running a model in their head. He states that the agent-based model is at least explicit, its assumptions transparent, and it can be validated. On the question of whether the agent-based model will be good at prediction, Epstein argues that we should instead focus on the many other reasons for modelling, of which he lists 18, including explaining certain phenomena, illuminating specific dynamics, discovering new questions, guiding new data collection, and disciplining the policy dialogue. Edmonds (2017) follows in Epstein's footsteps, exploring some of the major modelling purposes in greater detail, including their risks and mitigating measures. Edmonds explores prediction, explanation, theoretical exposition, description, and illustration, as well as some interconnections between these purposes.

Despite there clearly being many good reasons for modelling other than predication, the discussion about using agent-based models for prediction is far from resolved. Prediction remains (one of) the ultimate aim(s) of scientific research and pointing at alternative purposes is often seen as exposing a weakness in agent-based modelling. Elsenbroich and Polhill (2023) is a collection of papers presenting the diversity of positions on agent-based modelling and prediction. The problem with using agent-based models for prediction is almost intrinsic to their purpose. Complex systems are inherently unpredictable due to non-linearity, emergence, and adaptation. Agent-based models are there to better understand these features of particular complex systems modelled. The process of building an agent-based model, for example all those decisions we discussed in this chapter, the idealisations and abstractions, necessitate that we are not replicating the full complex system. Each boundary decision we take leaves the possibility open that we have left out a salient feature that will influence the future state of the real-world target system. Unless we acknowledge this we will do agent-based models a disservice.

Elsenbroich and Badham (2023) assert that while agent-based models are not themselves good to generate predictions, they help you understand the processes of change and will help make better decisions through this enhanced understanding. Picking up the framework of *justified stories* developed in Badham et al. (2021), we argue that agent-based modelling is a method to generate empirically grounded (justified) unfolding dynamics (stories) of social phenomena. This framework shifts the focus in the interpretation of an agent-based model from a specific output, a prediction or forecast, to the development and understanding of possible and plausible scenarios (for more details see Section 15.4).

2.6 Summary

This chapter introduced some of the key concepts of agent-based modelling: agents, environment, interactions etc. and, most importantly, the need to focus on modelling processes. We discussed the quagmire of modelling decisions that need to be made for an agent-based model, including deciding the boundary of the system modelled, questions of abstraction and idealisation, and the difficulty of operationalisation. Finally, we discussed questions about the purpose of agent-based modelling, as generative social science, moving away from traditional scientific conceptions of causality and the many different purposes of agent-based models other than predicting and forecasting.

Given this overarching paradigm of agent-based modelling, it should be seen as a methodology rather than just a method. The modelling approach is not a single tool but is a new way of thinking about society as a complex system that needs to be tackled by taking seriously the agents, actions, interactions, and emerging patterns observed in the real world. Conceptualising agent-based modelling as a methodology can also explain the great variety of agent-based models that can be built. We have variously mentioned different kinds of agent-based models, different levels of abstraction, complexity, and realism. While they differ in application and purpose, they all have in common that they replicate essential processes to better understand real-world patterns, all built on the foundation of generative causality.

Questions

1. What is an agent-based model?
2. List three things you have to specify when building an agent-based model.
3. Explain the difference between abstraction and idealisation.
4. Name three purposes of agent-based modelling other than prediction.
5. What is generative social science?

3

Ethics of Agent-Based Modelling

Ethics lies at the heart of the social sciences. Although agent-based modelling often does not involve human subjects directly, ethical considerations pervade every step of the modelling process. This chapter explores general social science ethical principles, relates them to agent-based modelling, reflects on the ethical stance of scientists, and concludes by examining questions of power dynamics in modelling.

3.1 Research Ethics

Ethics is the study of principles and values that guide right choices and behaviours. Research ethics specifically develops principles and values for research practices.

Research Ethics

Research ethics govern the principles and standards of conduct for scientific research.

Agent-based modelling has seen relatively little consideration of ethics until recently. Anzola et al. (2022) focuses on shifting ethics considerations from avoiding misconduct toward actively pursuing responsible conduct and research integrity. With agent-based modelling increasingly focusing on policy-relevant models, they argue it is time for the social simulation community to agree on and follow a code of ethics (see Figure 3.1).

But let us first look at ethics in the social sciences to have some context for discussing ethics in agent-based modelling.

3.2 Ethics in the Social Sciences

Ethics forms an essential part of any social science – indeed, any science concerned with sentient beings. Science must maximise utility and minimise suffering. However, ethics in social sciences only developed seriously from the mid-20th century.

The Holocaust's horrors, which included human experimentation without subject consideration, spurred the scientific community into action. It began with the Nuremberg Code in 1947, establishing protections for human subjects against misconduct and abuse of vulnerable persons. The Declaration of Helsinki and Belmont Report followed in 1964 and 1968, respectively, expanding on the rather rudimentary Nuremberg Code. From these underlying principles, several ethical considerations must underpin all social science research:

1. **Informed Consent** Any engagement and data collection with human subjects requires informed consent. Researchers must provide complete and clear information about the study's purpose, methods, potential risks, and benefits to participants and obtain their written consent. Participants give consent freely, without coercion, and should always be free to leave a study. Importantly, data must only be used for purposes stated in participant information and consent forms. Informed consent can become tricky with online data; though social media is in the public domain, using individual posts represents a grey area.
2. **Privacy and Confidentiality** Personal data privacy and confidentiality must be safeguarded. This means data must be anonymised and securely stored to prevent identification or breaches. Researchers must follow data protection laws and institutional guidelines, such as Europe's GDPR (General Data Protection Regulation).
3. **Avoiding Harm** Avoiding harm is central to scientific ethics. Participants should not face unnecessary physical, psychological, or social risks. This is particularly important when working with vulnerable groups like children or sensitive topics like trauma. Harm avoidance also applies to researchers, requiring risk assessment and mitigation before research begins.
4. **Avoiding Bias and Discrimination** Research must be as unbiased as possible. All participants deserve equal treatment without biases related to gender, race, socioeconomic status, or other attributes. This is especially important for generalisable research, e.g., quantitative studies, but the onus of non-discrimination is salient in all research. Researchers must ensure chosen methods are transparent to all participants to maintain neutrality and recognise shared understanding as the basis of knowledge.
5. **Honesty and Integrity** Researchers must accurately report findings, avoiding fabrication, falsification, or misrepresentation of data. This transparency ensures research integrity. All results should be presented, including negative and inconvenient findings (which can be difficult when working with stakeholders, as discussed in Section 3.4). Researchers must also avoid plagiarism and acknowledge all contributors in publications.
6. **Accountability and Governance** To demonstrate adherence to these principles, proper research governance is essential. This means maintaining records of every reseach step, including consent, ethics clearance, and data management.

3.3 Ethics of Agent-based Modelling

Adhering to ethical principles protects the dignity, rights, and welfare of researchers, research participants, and the public as a whole. Discussions about ethics in agent-based modelling are relatively recent. This is not because the field has been conducted unethically; rather, traditional research ethics focuses primarily on data collection and participant rights, which have not been central concerns in agent-based modelling. This has changed, however, with the advent of empirical agent-based models that can use extensive amounts of data and may be used for policy decision support.

As a social science method, agent-based modelling requires ethical considerations in the research process. Empirical agent-based models use social data – data about human beings. Standard questions of anonymity, identifiability, and informed consent apply. This becomes particularly important in agent-based modelling as different data types are often linked to create a more complete picture. However, data linkage can result in identifying persons not identifiable in unlinked datasets. Additionally, modellers might need to question the data itself, examining potential sampling biases, construct validity of questions, and approaches to missing data. Data-related ethics questions are discussed in Section 3.3.1, along with ethical dimensions of the model itself (Section 3.3.2) and the interpretation and use of modelling (Section 3.3.3).

An early code of conduct for modelling and simulation is presented in Oren et al. (2002). It focuses on the profession of modellers and simulationists and the ethical necessities of models for decision support. This code has been adopted widely by different modelling and simulation organisations. Tolk (2017) discusses the uptake and usefulness of the code, pointing in particular at the practical implications for modellers as scholars, working with stakeholders, and as agents in the modelling process, which Tolk sums up as follows:

> Modeling is the task-driven, purposeful simplification and abstraction of a perception of reality that is shaped by physical, ethical and cognitive constraints. (Tolk, 2017, p.47)

This view of modelling emphasises the normative origin of a model in both the *purpose* of the model and the assertion that the model is a representation of a *perception* of reality.

Tolk concludes that ethical conduct is critical not only for individual integrity but also for the credibility and advancement of the discipline of modelling and simulation as a whole. Professionals must be vigilant against bias, uphold transparency, and foster a supportive, inclusive environment, especially those in leadership positions.

Shults and Wildman (2019) evaluate three aspects of ethics in modelling and social simulation. The first one is the practice of social simulation and is largely tied to professional conduct, following Tolk. They reiterate the importance of transparency, highlighting the power computational models can have to convince people of certain realities.

A second aspect they discuss is modelling ethics directly, i.e. using models as scientific instruments to investigate ethics, for example using game theory to demonstrate the emergence of cooperation over time (starting with Axelrod, 1997, this has developed into an extensive research area). The authors argue that models and simulations help us to better understand ethical decision-making and policy planning, not only by exploring outcomes more thoroughly but also by clarifying the value assumptions behind decisions. Models can raise deeper philosophical questions about how we evaluate ethical outcomes and the purposes behind our modelling efforts.

The third aspect Shults and Wildman (2019) discuss is meta-ethics. While they acknowledge that science cannot resolve ultimate moral questions, it shapes and constrains moral discourse through data, making supernatural explanations appear less plausible. Simulation modelling can be used as such a constraint by demonstrating potential processes that can lead to the phenomena actually observed in the world (e.g. diversity of species through mutation and natural selection).

While Oren et al. (2002) and Tolk (2017) grapple more generally with modelling and simulation as an engineering pursuit, most of Shults and Wildman's (2019) examples are focused on social simulation.

The Code

Institutional considerations

1. **Professional competence and training**: Have in place plans for training for new and experienced modellers, not only on technical aspects of modelling, but also on model use and interpretation, and interdisciplinary working.
2. **Interdisciplinary working**: Have in place plans and common understandings of how the challenges of interdisciplinary working will be managed (e.g., different assumptions about the aims and value of methods and projects, knowledge asymmetries, different ethical standards).

Individual project considerations

1. **Project management, transparency, and quality**: Plan and maintain project management processes to ensure the documentation, quality, and reproducibility of all model stages.
2. **Narrative and positions**: Be honest and open about your underlying beliefs about the system you are modelling. Are there contested understandings, results, or interventions that you expect your model to support or contradict?
3. **Model use**: Develop a plan for how a model will be used and for unplanned use of your model or its results.
4. **Model inputs**: Be conscious of bias in the data and theory you use to inform a model.
5. **Model assumptions**: Be honest and transparent about the assumptions in your model and their implications.
6. **Stakeholders**: Involve or engage with any relevant stakeholders, including the people or organisations you are modelling.
7. **Representation**: Consider how people and/or groups are represented in your model; ensure no one or group is represented in an inappropriate or inequitable way.
8. **Open science**: Make all model code and analysis open and provide documentation
9. **Dissemination**: Consider the appropriate use of standards in model reporting.
10. **Publication**: Publish in an open and honest way, acknowledging failures and false starts.
11. **Authorship**: Ensure authorship and contributions are fairly attributed.
12. **Public communication**: Consider preparing documentation which presents the model, results, and their interpretation for layman or non-modeller audiences.
13. **Maintenance**: Make appropriate arrangements for long-term model and documentation maintenance.

FIGURE 3.1: Proposed draft of a code of ethics for agent-based social simulation (Anzola et al., 2022).

Anzola et al. (2022) is particularly relevant, explicitly examining the practice of agent-based modelling. They propose specific questions modellers should address at each step of the modelling process, and developing a proposed code of ethics (see Fig 3.1).

These emerging recognitions of ethics' importance in agent-based modelling are finding practical applications, such as the inclusion of ethics questions in peer reviews of agent-based modelling papers (Polhill et al., 2025). We very much encourage you to engage with ethical questions throughout your modelling project.

3.3.1 Ethics of Data

Most computational modelling uses secondary data, which does not raise the same ethical issues as primary data collection. However, researchers using secondary data must navigate several ethical considerations to ensure responsible, transparent, and respectful work.

Informed consent remains important, as original data collection might not have obtained permission for data reuse. Secondary data use can compromise personal data protection and anonymity. Data might not be properly anonymised, risking individual identification, particularly with small samples or rich datasets.

Linking datasets can enable cross-referencing, potentially leading to re-identification. This poses a particular danger for agent-based modelling, which often combines different datasets to create rich empirical foundations. Special care is needed with data involving children, marginalised populations, or other vulnerable groups.

Secondary data use also risks misinterpreting data if the original research context is not fully understood. This can lead to misapplying data to support preconceived ideas and distort research integrity, a particular danger in agent-based modelling due to the potentially idiosyncratic nature of model building. Selective and decontextualised data use can misrepresent communities or individuals of cultural and social groups, requiring care to avoid distortion and stereotyping. Another danger lies in building models driven by data availability rather than research questions; although data availability matters, it should never drive model development.

When selecting data for agent-based modelling, researchers must ensure data integrity and validity. They must critically assess the quality and credibility of secondary data sources. Transparent reporting of data limitations, methodologies, and potential biases is crucial. Researchers must also adhere to licensing agreements and legal permissions for data use; data should never be used without proper authorisation or by bypassing access restrictions.

As in all research, transparent reporting is key to ethical data use in agent-based modelling. This means clearly stating what data were used, how they were interpreted, and properly citing original data sources.

3.3.2 Ethics of Modelling

Building a model might seem innocent: a few assumptions, some data tossed into code, resulting in a story. But models can become powerful tools with real-world consequences, and if used wrongly or unethically these consequences can be dire. For example a model of evacuation of a building in case of a fire that does not take into account wheelchair users might lead to catastrophic building decisions. In this section we focus on important ethical considerations when building the model itself.

Agent-based models can effectively represent social phenomena but risk oversimplification. This could mean excluding particular population groups but could also involve representation choices about structural features or behaviours. For example, Schelling's segregation model helps us understand segregation dynamics but cannot fully explain the complex realities of segregation in USA cities, which involve racism, migration, and socioeconomic factors. Simplistic models might mislead by suggesting inequality or segregation as "natural states", especially if embedded in empirical data, making oversimplified dynamics seem validated. Researchers must clearly communicate their models' limitations and complexity to avoid ideological overreach.

Choices made in the conceptual model carry into subsequent model-building steps, including theory choice, data selection, operationalising concepts into computational models, and analysis. Reflecting on the ethics of these choices and reporting them openly and transparently is essential. Anzola et al. (2022) discusses ethical implications at each modelling step, most concerning cogency and transparency.

Anzola et al. (2022) also addresses the ethics of reporting agent-based models. Ethics of reporting includes open availability of publications and code. They also frame results reporting as ethical questions of whether results are published accessibly, fairly, and without bias. The question of accessible reporting is particularly interesting for agent-based modelling. Computational models can intimidate less technically savvy audiences, allowing outputs to be misused by appearing more convincing than warranted. Even without malicious intent, computational models can be misrepresented by focusing too much on particular outcomes (rather than the range of potential outcomes) or keeping reporting too mathematical. When reporting an agent-based model, consider visualisations and narratives that help lay people understand the model better. Remember, most people have limited knowledge of agent-based modelling.

3.3.3 Ethics of Model Use

Models have power. At the beginning of the COVID-19 pandemic, a model developed by researchers at Imperial College London (Ferguson et al., 2020) finally convinced UK Prime Minister Johnson to impose a lockdown to reduce virus spread. The model predicted around 500,000 UK deaths over two years without interventions. This individual-based model stipulated social contacts in households, schools, workplaces, and communities. Census data provided demographic calibration; schools and workplaces were distributed throughout the population according to population density and commuting distances. The model used a classic Susceptible-Exposed-Infected-Recovered (SEIR) underlying logic, where transmission occurs when susceptible and infectious individuals meet in any contact setting. Transmission rates were parameterised using past influenza data and social mixing surveys (Mossong et al., 2008).

This model's prominence in the UK government's lockdown decision led to heavy criticism of both the model and Neil Ferguson (along with many other scientists) personally. In May 2020, John Fund wrote in the National Review:[1]

> Last March, Ferguson admitted that his Imperial College model of the COVID-19 disease was based on undocumented, 13-year-old computer code that was intended to be used for a feared influenza pandemic, rather than a coronavirus. Ferguson declined to release his original code so other scientists could check his results. He only released a heavily revised set of code last week, after a six-week delay.

Both sides faced accusations of ideological bias — modellers favouring severe interventions, critics ideologically opposing interventions to reduce social contacts as excessive liberty restrictions. The quote criticises the Imperial College Model for using undocumented, 13-year-old code based on influenza and not releasing the original code.

Regarding modelling ethics, undocumented code and not releasing model code clearly violate the transparency requirements discussed earlier. Thirteen year old model code might be problematic if the model was oversimplified due to computational capacity restrictions but is not necessarily problematic. Unless the world changes, a model should remain applicable. The model was built for influenza rather than COVID-19, which could pose serious problems if relevant transmission differences existed between the modelled virus and the target system. *Relevant differences* means the virus transmits differently, such as the difference between an airborne disease and a sexually transmitted disease. While coronavirus transmission was not completely understood in March 2020, observed transmission patterns in places like Italy indicated sufficient similarity to influenza.

But should one model change history's course and confine an entire population to their homes? Of course not. Importantly, that is not what happened here. The model was one influential piece of evidence in a growing mountain of information showing disaster was imminent and action necessary. The model helped think through consequences, both death toll and, more importantly, intensive care provision and health service capacity. The lockdown decisions followed the modelling outputs together with other data, like the observation of Italy's nearly collapsed health system.

[1] [https://www.nationalreview.com/corner/professor-lockdown-modeler-resigns-in-disgrace/]

The episode highlights modelling problems and sparked ongoing debate about agent-based modelling's role in crisis support. Squazzoni et al. (2020) calls for agent-based modellers to become more involved in crisis modelling, promoting agent-based modelling as better for modelling crises due to its capacity to handle system complexities. This view is not unanimous across the community, though (see the thread following Elsenbroich and Badham (2020)).

The Imperial College model story raises questions about models as decision-makers rather than decision support tools. Computational models might be used unethically when their limitations are not adequately communicated, leading to biased or misleading results influencing significant decisions. Predictive policing uses past crime data to predict future crime and allocate police to potential hotspots but faces criticism for racial bias and over-policing minority communities (Ugwudike, 2022). Similar biases emerge from computational models for credit scoring that unfairly penalise certain demographics (Fuster et al., 2022), facial recognition technology leading to wrongful surveillance based on biased algorithms (Leslie, 2020), and health risk prediction models leading to biased treatment decisions (Paulus and Kent, 2020).

These examples come from data-driven computational modelling, machine learning, and big data but highlight problems also affecting agent-based modelling: potentially biased or incomplete data and lack of transparency. Agent-based modelling increasingly integrates with AI-driven technology, creating new potential ethical problems. Polhill et al. (2025) identifies three uses of AI and their ethical problems:

1. AI can create synthetic data where real data is scarce, opening models to the same bias problems discussed earlier.
2. Using Large Language Models in coding enhances efficiency by suggesting code, identifying errors, and optimising algorithms, but algorithm development becomes opaque to human modellers, potentially compromising model accountability and propagating errors or biases.
3. AI can analyse and interpret complex model outputs, offering summaries, insights, and visualisations. While efficiently handling large datasets, this risks oversimplification, misinterpretation, or bias propagation. Researchers might over-trust AI interpretations, potentially missing critical nuances that human analysts would catch.

Another important ethical problem is potential over-reliance on the predictive power of agent-based models without considering real-world system complexity. Agent-based models replicate complex systems, which inherently challenge prediction. Modeller decisions about model boundaries, abstractions, and idealisations introduce additional challenges for predicting real-world outcomes. Treating model outputs as definitive truths without considering uncertainties and limitations poses a major (ethical and practical!) problem. The complexity of agent-based models, combined with their intuitive appeal, creates an additional challenge as non-experts might not understand uncertainties and limitations, potentially enabling unethical applications without proper scrutiny.

3.4 Whose Model and Who's Modelled?

This section discusses two power relationship questions important to understand in all research, including agent-based modelling. The first concerns who the model is for and how this influences the model and modelling process. We discuss the stakeholder-modeller relationship. The second concerns who is included and excluded from an agent-based model, i.e. ethical questions about representation.

3.4.1 Whose Model?

One of the first questions when building an agent-based model is the audience the model is built for. Beyond implications for the model building process, this raises ethical considerations. Modelling for clients or stakeholders versus general social science research creates different ownership relationships, reporting questions, and representations. When working with clients and stakeholders you might face the problem that clients might dislike research findings. This problem potentially intensifies with agent-based models, as decisions about what to include can skew models in certain directions. Utmost care must be taken that modelling decisions are not led by vested interests.

We have seen above that models can be powerful tools to convince people that certain things will happen in the future. As agent-based models rely on assumptions, it is important to ensure that the assumptions are not led by what the researcher wants to demonstrate. For example, an ABM of health inequalities that assumes a lower propensity for poorer people to engage in physical exercise and has physical exercise as the driver of good health will "find" socio-economic health inequalities. But the result is "baked into" the dynamics of the model, making the model more of an animation than an instrument of scientific discovery.

When stakeholders include diverse groups (policy partners, commercial companies, lived experience participants), carefully consider how to represent different views in the model. Consider power relationships between stakeholders during research design, for example, whether participatory sessions should mix across stakeholders (allowing all voices to be heard simultaneously and discussion about different perspectives, but risking that power dynamics might silence some voices).

But also the modeller has a stake in the model. Edmonds (2023) discusses various experiences where modellers reported ethical concerns, including pressure to align results from different models despite varying assumptions, exaggerate predictive capabilities in grant applications, rush integration of findings within unrealistic timeframes, suppress negative results, and oversimplify conclusions, stripping them of essential complexities for decision-makers. In response, Edmonds (2023) developed a declaration of rights for modellers. One right, the right to accurate attribution, matters even when modelling for general social science. Sometimes code writing gets outsourced to programmers or software engineers. It is important to acknowledge *all* model contributions.

Another ethical consideration is what happens to models after publication. Good practice allows open access to models, including model code, enabling other modellers to inspect, replicate, and critique them. However, this also enables researchers to use, adapt, or extend your model, potentially opening possibilities for model misuse. While no known instances exist, it is easy to imagine for example misusing an evacuation model to target a terrorist attack for maximum casualties.

Finally, Polhill et al. (2025) makes an interesting point about equity in agent-based modelling as a policy tool. Some agent-based models require high-performance computing, creating challenges in low-income countries and risking dependency on foreign technology. Local knowledge is crucial for accuracy, and ethical concerns arise if simulations influence harmful policies amid power imbalances.

3.4.2 Who's Modelled?

The second power relationship question concerns who the model represents. A frequently highlighted strength of agent-based modelling is representing heterogeneity. Simultaneously, we make simplifying assumptions that exclude certain aspects, i.e. restricting the diversity we model. A previously used example of modelling the labour market without including gender has the potential to obscure gender pay gaps. Including gender dynamics for pay gaps might in turn obscure labour market disadvantages through race, ethnicity, or social class. This section examines who our models represent and who might be excluded, and why.

These questions begin during early modelling stages, formulating research questions, establishing model scope and purpose. We must consider who has input at this stage, who benefits from answering the research question, and who is considered relevant for inclusion in the model. Anzola et al. (2022) rightly asks:

> Do disempowered groups, or those that are the subject of analysis, or who may be affected by decisions arising from it, have a say?

As with many ethical questions, no right answer exists. However, it is important to remember that modelling decisions necessarily exclude some populations. Is it ethical in an evacuation simulation to exclude people with physical disabilities, such as wheelchair users? Representing all populations is impossible, but we should consider whether populations have been excluded or represented conveniently or unfairly. This likely will not result in explicitly modelling all sub-populations, but noting the exclusion should inform discussion about model limitations. As always with agent-based modelling, transparency about choices and justification of model-building decisions are crucial.

3.5 Summary

Research should always aim to contribute to knowledge and understanding and generate positive societal impact. Agent-based modelling is no different, meaning all normal social science research ethical constraints apply (general ethics and specific issues with secondary data).

Some specific ethical considerations exist for computational and agent-based modelling, mainly concerning honesty, transparency, and accessibility of models and modelling decisions. Model use creates its own ethical problems, particularly when models are used ideologically, without context, and when limitations are not clearly communicated.

A major question for agent-based modelling involves power relationships in the modelling process. Models are often commissioned by or built with stakeholders, raising questions about ideology and bias, releasing unwanted model outcomes, who benefits from the model, and demands placed on modellers.

The second important power relationship stems from the ability of agent-based models to represent heterogeneity. This core strength opens important questions about who is represented and who is not in an agent-based model. These decisions necessarily occur within the modelling process, but it is important to be conscious of them and report them as potential limitations.

Questions

1. What moral principles guide your research and research problem selection?
2. What potential ethical issues affect your research: the design, data, model design, and modelling purpose?
3. What responsibility do you have regarding communication of your modelling outputs?
4. Is your research beneficial for society as a whole?
5. Why might it matter who the model is built for?
6. Discuss why inclusion and exclusion questions are particularly important for agent-based modelling. Give an example.

Part II

Building Agent-based Models

4

Preliminaries

This chapter prepares you to start the tutorial. We formally introduce NetLogo, the software and language used. You will play with some library models and start to connect concepts already discussed with their implementation. Finally, we describe the model that is to be developed during the tutorial, which concerns protective behaviour (such as good hand hygiene) in response to an epidemic. Although we don't start coding yet, we think about some of the important model design questions.

4.1 Introducing NetLogo

Before starting to code the tutorial model, you must first install and become familiar with the modelling software *NetLogo*. The NetLogo software includes a library of example models, and we will use these to demonstrate the concepts already introduced, focusing on agent-centric thinking and complex system behaviour.

NetLogo (Wilensky, 1999a) is a specialist agent-based modelling environment that includes its own programming language. It can be downloaded from https://ccl.northwestern.edu/netlogo and is available (for free) for Windows, Mac OS X, and Linux. This book assumes you are using version 7 (the latest available at the time of writing).

There is also an online version of NetLogo (called NetLogo Web) that can be run over the internet, without installing any software. However, only some NetLogo keywords are included, and the tutorial cannot be completed over *NetLogo Web*.

Download and install NetLogo.

4.1.1 Navigating a NetLogo model

Open NetLogo. As no model is loaded, an empty model template is displayed (see Figure 4.1). The main menu at the top (labelled 1 in the figure) could be in a slightly different position depending on your operating system. This menu is used to open an existing model, save the model you are working on (both in the File menu), and run experiments (with the batch simulation tool BehaviorSpace; see chapter 13). It also provides access to resources in the Help menu such as the NetLogo Dictionary, which describes all the keywords.

Below the menu, there are three tabs in the centre of the interface to switch between the 'Interface', 'Info', and 'Code' screens (labelled 3 in the figure). The interface tab is the visible screen when NetLogo starts. When a model is open, the interface contains the buttons, sliders, plots, and other model controls for the user to operate the model and interpret its results. The

Info page contains descriptive information about the open model. Models in the sample library included with the NetLogo software have useful information similar to a basic user guide for the model. As this information is written by the model developer, detail may be limited for other models. The Code tab accesses the page with the program code that implements the model.

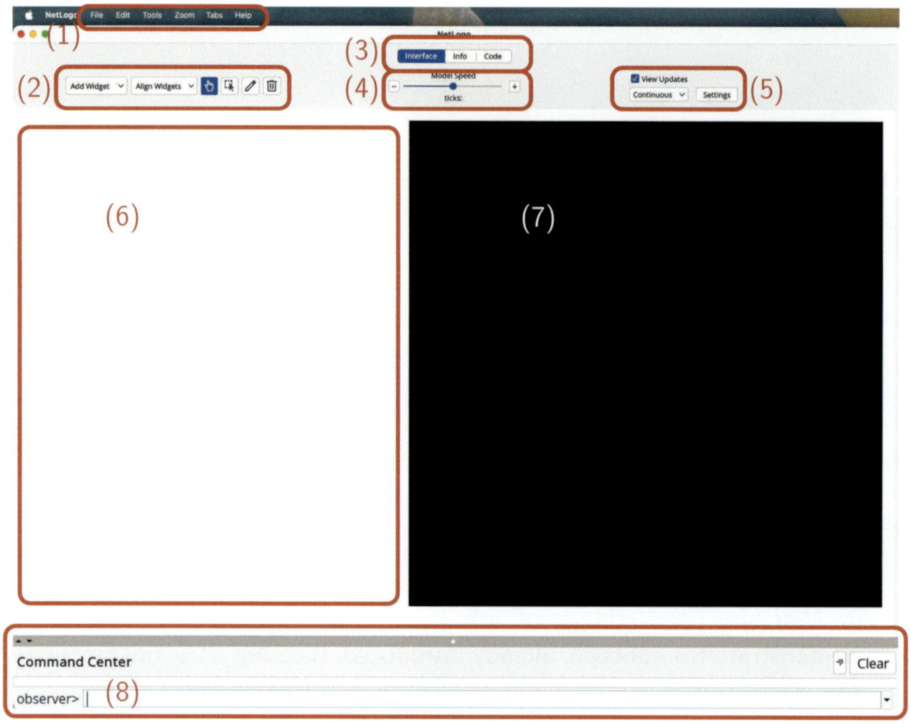

FIGURE 4.1: Screenshot of an empty NetLogo model with parts of the interface labelled. (1) is the menu. (2) adds widgets (input controls and output reporters) to the interface and controls their placement. (3) is the tabs to switch between the Interface, Info and Code views. (4) contains the speed slider. (5) is update frequency and settings controls for the world view. (6) is a blank area of the interface where input and output widgets are added. (7) is the world, or main view of the model. (8) is the command center, where commands can be typed in to run code interactively.

The functions of the other areas will be described as they are needed. Briefly, the model developer uses the widget (2) and settings (5) items to design the interface for the user to interact with the model. The main interface contains a view of the world (7), various input and output widgets to control and interpret the model (in the currently blank area 6), and controls for the simulation speed (4). Finally, the Command Center (8) allows the developer or the user to directly enter NetLogo code for immediate running.

Widget

A widget is an interface component such as a button, plot, or slider.

4.1.2 NetLogo model entities

For historical reasons, the main agents in a NetLogo model are referred to as *turtles*.[1] These are the agents that perceive their spatial and social environment and change their behaviour accordingly. In this tutorial, the term 'agent' and 'turtle' will be used interchangeably, except where otherwise stated.

There are three other types of entities in a NetLogo model. The spatial environment comprises *patches* and the social environment is comprised of *links* that represent a relationship between two turtles. More abstractly, NetLogo is aware of the *observer*, who is able to issue instructions to the other entities.

Technically, all of these entities are different types of agents within the architecture of the NetLogo software. However, referring to *patches* and *links* as agents is unnecessarily confusing for novices because they are conceptually different from the agents whose behaviour is actually being simulated. Nevertheless, it is important to be aware of this usage because the official documentation refers to all of the entity types as agents.

4.2 Understanding NetLogo Models

The next step in NetLogo familiarisation is to open (load into NetLogo) some existing models and play with them. There is a library of sample models installed with the software, which contains full models and short examples of code to achieve specific results. Open the *Models Library*. This can be found as an item in the *File* menu (see Figure 4.2).

4.2.1 Rabbits Grass Weeds

From the library, load the model named *Rabbits Grass Weeds* (located in the Biology section of the Sample Models section, or search for 'rabbits' using the search box at the bottom of the models list). When initially opened, the interface has widgets on the left (in the area marked (6) in Figure 4.1) but the world (area (7)) remains empty. Once the model is running, it will look like Figure 4.3.

The user controls consist of two buttons (labelled 'setup' and 'go') and six sliders with various labels (section (1) of Figure 4.3). In addition, there are two output widgets, a plot labelled 'Populations' and a small box labelled 'count rabbits' (section (3) of Figure 4.3).

Drag the 'number' slider to set the initial number of rabbits. Do this by placing the mouse pointer on the white circle at the end of the blue bar and hold down the mouse button to move the circle. You can also change the value a small amount by clicking the mouse cursor to the right or left of the circle. Press (with a mouse click) the button labelled 'setup'. The 'setup' button is used to initialise the simulation, creating some patches with grass and a starting population of rabbits. The world should now have small green squares and white rabbit icons scattered (numbered (2) in Figure 4.3).

[1]The Logo programming language was used in the 1980s in education settings to program robot turtles to draw by crawling on a piece of paper with a pen. Creating a pattern requires taking the turtle's perspective instead of an observer's.

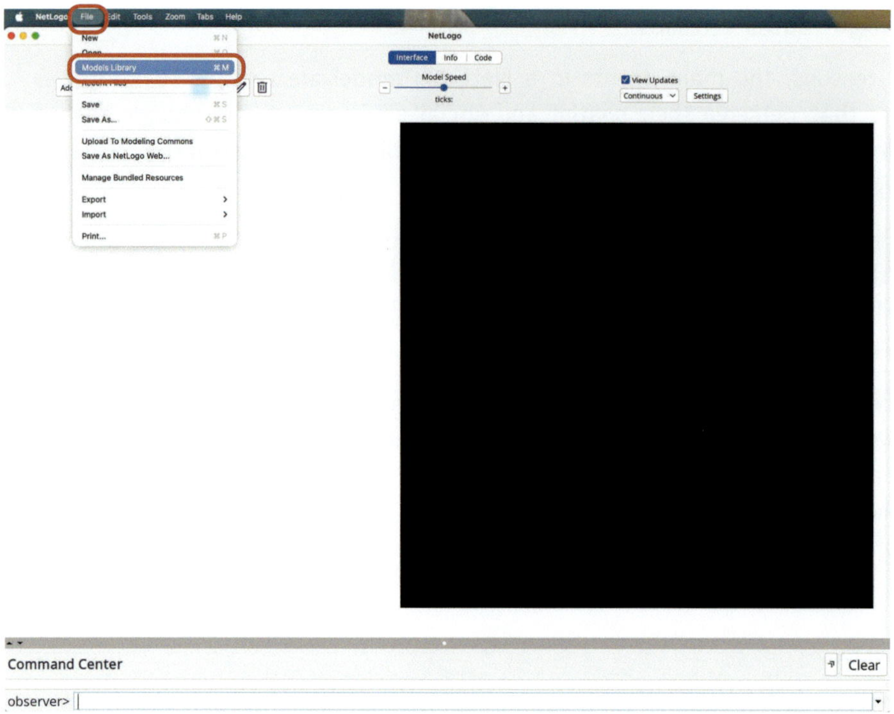

FIGURE 4.2: Screenshot of an empty NetLogo model. The menu option for the models library is highlighted.

Start the simulation running by pressing the 'go' button. To pause (or stop) the simulation, press the 'go' button again and then continue the run with the same button.

Once the simulation is running, the plot will display the number of patches with grass (green) and rabbits through time. The current number of rabbits is in the monitor (small box below the plot). Use the speed slider (numbered (4) in Figure 4.3) to speed up or slow down the simulation.

Play with the model for at least several minutes. Try out different values for the sliders. Remember to press the 'setup' button to initialise a new simulation and then the 'go' button to start it running. Here's some things you should try, and some questions to consider while playing:

- How would you describe the system level behaviour to someone who can't see the model running? For example, is there a pattern over time for the rabbit population? Does there appear to be a relationship between the different populations (curves in the plot)?
- What happens if you move the slider labelled 'grass-growth-rate'?
- What about the other sliders? Try moving several sliders at once and then just one slider at a time.
- While the simulation has been running for a while, move the 'birth-threshold' slider while the model is running.
- Can you work out what is happening within the model just by playing with it?

Do not read past this point until you have completed playing with the model. The next paragraphs discuss some of these experiences, and you will not learn as much if you simply read the discussion.

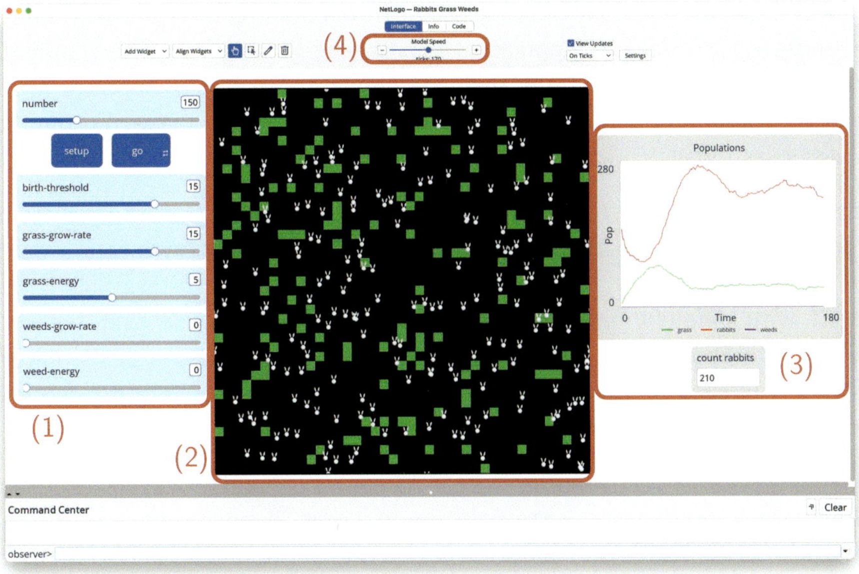

FIGURE 4.3: Interface for the *Rabbits Grass Weeds* model part way through a simulation run. In this model, the input controls are in the left section of the interface (1), the world (2) displays the location of rabbits and patches of grass, and key information about the simulation is reported to the right (3). The speed slider (4) controls how fast the simulation runs.

The next step is to look at the Info tab. This contains text written by the modeller. As *Rabbits Grass Weeds* is a library model included with the NetLogo software, the text follows a specific format that is geared toward eduction. It includes a brief description of the behaviour mechanisms within the model, what to adjust, and what to look for. There is also some information about aspects of the NetLogo language that are particularly interesting. The language discussion can be ignored at this stage and will be revisited at the appropriate point in the tutorial.

4.2.1.1 Complex behaviour

Rabbits Grass Weeds is a predator-prey model, an important type of model in ecology. In this case, rabbits move around the world and eat any grass or weeds at patches they land on; rabbits are the predators and grass (and weeds) is the prey. These can be modelled mathematically with the Lotka-Volterra equations (Lotka, 1925; Volterra, 1926), but a simulation is easier to explore.

Predator-prey models have linked oscillating cycles of population counts. There is feedback between the two populations. When there are few rabbits, the grass is able to grow. But as there are more grass patches available, the rabbits are able to acquire energy and reproduce. With more rabbits, there is over-exploitation of the grass and a collapse in the rabbit population, starting the population cycle again.

The sliders have meaningful names, so it is likely that you understood at least some of what was happening within the model just from playing with it and thinking about what happens under different conditions. The mechanics of grass growth and rabbit reproduction are relatively straightforward: as the rabbit finds food, it gains energy from that food and then reproduces if its energy is sufficiently high (birth-threshold). What is less clear is that simply being alive costs a small amount of energy and that rabbits die when they run out of energy.

The rabbit population pattern can only be understood through its relationship with the quantity of grass (and conversely). The linked population curves emerge from the interactions between rabbits and grass. It is possible to observe that the curves move in the opposite direction of each other, and that the grass curve is slightly later than the rabbits' curve. Furthermore, the oscillations stabilise over time.

There are several other observable relationships. These are easier to extract by systematically moving one slider and holding the others constant. Return the weed-growth-rate slider to 0 (if you have moved it off 0) and set grass-growth-rate, grass-energy, and birth-threshold to moderate values. Then start with different numbers of rabbits. You should notice that the simulation eventually stabilises with about the same number of rabbits, but the shapes of the initial curves are different. This sort of exploration is easy to accomplish with a model, but not feasible in the real-world system that the model is intended to represent.

One aspect of complex systems that you may have experienced is the difficulty in theorising about the micro-level behaviour mechanisms when only the macro-level system behaviour can be observed. Without a clear understanding of causes, it is also difficult to understand the consequences of intervening in the system.

Imagine that you had observed a short period of the oscillating rabbit population in the real world. Many potential interpretations are possible including some unknown predator, a disease that affects fertility, and changes in food quality. Reductionist research techniques such as examining individual rabbits will not help to understand what is happening because those techniques do not examine context and interactions.

The difficulty is exacerbated where there are multiple simultaneous changes. Experimenting with the model allowed you to isolate the effect of individual changes and simplify the behaviour (for example, by excluding weeds with weed-growth-rate set to 0). You also had the benefit of informative variable names. None of this is true when trying to understand real-world complex behaviour.

4.2.1.2 Agent-centric code

To clarify the concept of agent-centric thinking, we need to look at the programming code. Press the Code tab (in area (3) of Figure 4.1). What you see is the NetLogo program that implements the *Rabbits Grass Weeds* model. You are not expected to understand any of the code at this stage.

You will see that there are several words that are coloured. These coloured words are the keywords in the NetLogo programming language that tell the computer what to do. In NetLogo, many keywords are standard English words such as 'ask'.

Keyword

A keyword in a computer programming language is a word that is part of the language and has a special meaning. Keywords allow the programmer to tell the computer what to do. The programmer cannot use those words for other purposes (such as variable names), they are reserved for their specified meaning. In NetLogo, the keywords that operate on agents and other entities (rather than the model as a whole) are also referred to as primitives.

The scroll bar at the right allows you to move through the entire program. The drop-down box labelled 'Procedures' allows you to move to specific sections of the code. It may be surprising to you that the program is very short to achieve so many things, including: grow grass, have rabbits be born and die, respond to the user's inputs, and visualise the results.

The 'go' procedure is displayed in Figure 4.4. This is the procedure that is repeated while the model is running. In the middle of this procedure is the command to 'ask rabbits' to do a list of five things delimited by square brackets. Each of these five tasks is actually the name of a separate procedure (which you can see because all the names are in the 'Procedures' drop-down box). The first of these is named 'move' and is displayed in Figure 4.5 and the second is 'eat-grass' (Figure 4.6).

```
to go
  if not any? rabbits [ stop ]
  grow-grass-and-weeds
  ask rabbits
  [ move
    eat-grass
    eat-weeds
    reproduce
    death ]
  tick
end
```

FIGURE 4.4: The 'go' procedure of the *Rabbits Grass Weeds* model contains the main structure of the program. Part of this code instructs each rabbit to do the five actions described in the procedures named in the square brackets.

Looking first at the 'move' procedure, this code has the rabbit turn a small random amount to the right and left, thereby ending up facing in a similar but not identical direction as it started, and then move forward. The rabbit's energy is also reduced slightly.

That is, the 'go' procedure instructs the computer to focus on one rabbit and have it do five tasks. First, the rabbit follows the instructions in the 'move' procedure, to slightly change direction, go forward a little, and consume some of its energy store. The next two tasks instruct the rabbit to look at where it is standing and, if there is grass or weeds available, eat the grass or weeds and obtain the energy that it holds (see Figure 4.6). If the rabbit has acquired enough energy, it births another rabbit (not shown). But if there was no food and its energy runs out, the rabbit dies (not shown). After running through all those actions for one rabbit, the program changes focus to the next rabbit until all rabbits have completed the actions.

This group of procedures demonstrates several aspects of agent-centric thinking, and how to implement such a perspective in a programming language. The 'go' procedure explicitly changes the focus to each rabbit in turn (using `ask rabbits`) and the model is implemented through the actions of the rabbits. The rabbit is able to perceive its own environment, in particular whether there is grass or weeds to eat. If so, the rabbit eats that grass or weeds and increases its own energy. The rabbit's action changes its environment, eating the weeds or grass means that any

```
to move  ;; rabbit procedure
  rt random 50
  lt random 50
  fd 1
  ;; moving takes some energy
  set energy energy - 0.5
end
```

FIGURE 4.5: The 'move' procedure of the *Rabbits Grass Weeds* model contains the first action that rabbits are instructed to follow in the 'go' procedure. First the rabbit performs a right turn ('rt') of a randomly chosen amount up to 50 degrees, then a similar left turn ('lt'). The rabbit then moves forward ('fd') a distance of 1 unit. Its energy is reduced by the amount 0.5.

```
to eat-grass  ;; rabbit procedure
  ;; gain "grass-energy" by eating grass
  if pcolor = green
  [ set pcolor black
    set energy energy + grass-energy ]
end
```

FIGURE 4.6: The 'eat-grass' procedure of the *Rabbits Grass Weeds* model contains the second action that rabbits are instructed to follow in the 'go' procedure. The rabbit looks at the colour of the patch where it is standing ('pcolor') and, if it is green (indicating grass), then the patch colour is changed to black and the rabbit's energy is increased by the amount of energy contained in grass.

other rabbits on the same patch are not able to obtain food. Therefore, although each rabbit operates independently, the behaviour of each rabbit affects the others and there is indirect interaction between rabbits as well as directly between rabbits and their environment.

As an aside, the code extracts also demonstrate the use of comments. The semicolon indicates to NetLogo to ignore the rest of the line (such as the ';' at the beginning of the second line in Figure 4.6). The programmer includes comments to assist human readers to understand what the code is intended to achieve. Comments are discussed in more detail within the tutorial (see Section 5.2.2).

4.2.2 Virus on a Network

The next model to explore is *Virus on a Network*. Open the *Models Library* from the *File* menu (see Figure 4.2) and load the model. It is in the Networks section of the Sample Models section, or search for 'virus' using the search box at the bottom of the models list. The *Rabbits Grass Weeds* used a spatial environment as the rabbits moved around a world of patches with grass. In contrast, this model demonstrates a social environment, represented by links between agents.

After pressing the 'setup' button, a network is created. Each node (small filled circle) represents an individual and edges (lines) represent a relationship between pairs of individuals. At the start, some of the nodes are infected and all others are susceptible to infection. Infected nodes have a probability of infecting any susceptible nodes to which they are connected. Once a node is infected, there is a probability of recovering each time step, either becoming susceptible again or immune. In epidemiology, these models are referred to as SIS or SIR models, but are both examples of the broader class of diffusion models. The names denote the available states; SIS for susceptible, infected, susceptible, and SIR for susceptible, infected, removed (indicating death or immunity).

There are two important sliders to explore the system behaviour. The slider named 'average-node-degree' varies the number of edges for each node (interesting values are about 10 to 30). The slider named 'gain-resistance-chance' sets the probability that a node will become immune to further infections when it recovers from its first infection.

Play with the model. Remember to press the 'setup' button to initialise a new simulation and then the 'go' button to start it running. Focus particularly on the how you would describe the system behaviour, and how it varies in response to changes in the slider values. Once you have experimented a little, read the Info tab to get some additional understanding of the model and further ideas about what to explore.

4.2.2.1 Agent-centric code

The 'spread-virus' procedure (Figure 4.7) implements transmission of the infection along network edges. The program selects one of the nodes that is currently infected. It then identifies all nodes that have an edge with that node. If any of these nodes are susceptible, they independently become infected with probability controlled by a slider on the interface (named 'virus-spread-chance').

```
to spread-virus
  ask turtles with [infected?]
    [ ask link-neighbors with [not resistant?]
        [ if random-float 100 < virus-spread-chance
            [ become-infected ] ] ]
end
```

FIGURE 4.7: The 'spread-virus' procedure of the *Virus on a Network* model is the main procedure called by the 'go' procedure. Infected network nodes (turtles) check the nodes at the other end of each of the links to which they are attached and, if the node is not resistant, there is a probability of infecting that node.

The actions entailed in becoming infected are implemented in the 'become-infected' procedure (Figure 4.8). The node's infected status is set to 'true' so that it will be included in the set of infected nodes in the next iteration of the 'spread-virus' procedure. The node is also coloured red for the convenience of the model user, providing a visual representation of the spread of the virus.

```
to become-infected  ;; turtle procedure
  set infected? true
  set resistant? false
  set color red
end
```

FIGURE 4.8: The 'become-infected' procedure of the *Virus on a Network* model is called by the 'spread-virus' for those nodes that are to be infected. It changes the colour and some status indicators of the node.

4.2.2.2 Complex behaviour

In the *Rabbits Grass Weeds* model, the rabbits interacted only indirectly; a rabbit's action in eating grass made that grass unavailable for other rabbits. In the *Virus on a Network* model, however, interaction is direct. A node transmits the infection directly to another node with which it has a relationship (conceptually, along the line or edge linking them). This interaction is the micro-scale mechanism that generates a system level behaviour, that of an epidemic. An epidemic is characterised as widespread infection within a population, and can be seen in the interface plots by an increase then decrease in the infected (red) proportion of nodes. That is, an epidemic is an example of complex system behaviour.

In the classic SIR model of differential equations (Kermack and McKendrick, 1927), the infection is spread by contact but a key assumption is that everyone has an equal chance of coming into contact with anyone else. That is, social networks are ignored. The initially infected people are embedded in a population of susceptible people and are able to transmit with every interaction. Consider the case where, on average, a person is infected long enough to transmit to two other people. This means that initially the infection spreads fast, doubling the new infections each generation. Eventually, however, infected people are coming into contact with people who are already infected (or recovered and immune), so the spread slows down. This sequence generates the classic diffusion curve that is observed in the *Virus on a Network* plots, where the number of nodes currently infected initially increases and then decreases. If the number of nodes ever infected is plotted, it would display an S-shaped curve. This is characteristic of diffusion models generally, such as adoption of a new product or other innovation (Bass, 1969; Rogers, 2003).

An SIR epidemic on a network shows similar behaviour, but the transmission is attenuated. As the infection can only be transmitted along network edges, the area of the network where the epidemic is active has a relatively high proportion of nodes that are already infected or immune. Slow the NetLogo model (using the speed slider, numbered (4) in Figure 4.1) to observe the transmission and try to identify areas where the epidemic is being blocked by an immune node. Blocking of the epidemic is more visible (and more extreme) with a lower number of edges per node because there are fewer options for the infection to access other areas of the network.

4.2.3 More Library Models

Playing with these two models was intended to familiarise you with the NetLogo interface, particularly buttons, sliders, plots, and the world display. In addition, you should understand how to access the Code tab and return to the Interface. Those models were selected because they demonstrate NetLogo physical and social environments. In both, the agents (rabbits and nodes) interact with their environment and take some action accordingly. That action can be a decision, such as eating grass, or simply a reactive status change, such as becoming infected. The sequence of actions is the simulated process that generates the complex system behaviour.

The Models Library contains many other models, demonstrating a wide variety of complex system behaviours. Several were discussed in Section 1.5 to introduce different ways in which the decisions of individuals generate some social phenomenon. Three additional models are briefly discussed here to demonstrate diverse processes, concentrating on the agent-centric perspective and how that generates system behaviour. For these, it is sufficient to read the Info tab and run the model with a few different settings; there is no need to examine the code.

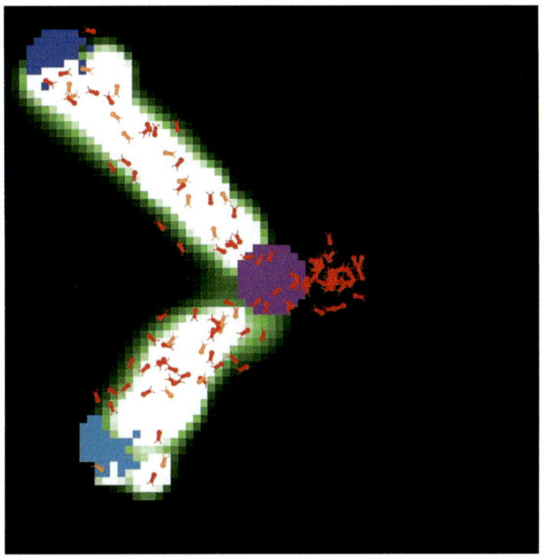

FIGURE 4.9: Ants model

The *Ants* model (in the Biology section) implements a process where each ant interacts only with its own local environment but the entire colony efficiently finds food and returns it to the nest. The ants explore randomly until one (or more) finds food. When an ant finds food, it brings the food back to the nest. While carrying food, the ant leaves a pheromone trail, which gradually spreads out and evaporates (Figure 4.9). Ants that cross a pheromone trail move in the direction of the strongest pheromone and then reinforce that trail when they find the food. That is, the ants communicate locally with pheromones to inform each other where the food is located.

The *Ethnocentrism* model (in the Social Science section) is related to the *Prisoner's Dilemma* model discussed at Section 1.5.2. Evolutionary game theory explores the effect at the population level of repeated interactions between individuals who are able to assist each other at some personal cost. They were initially developed to explain how altruism could arise despite its evolutionary disadvantage (Smith, 1964).

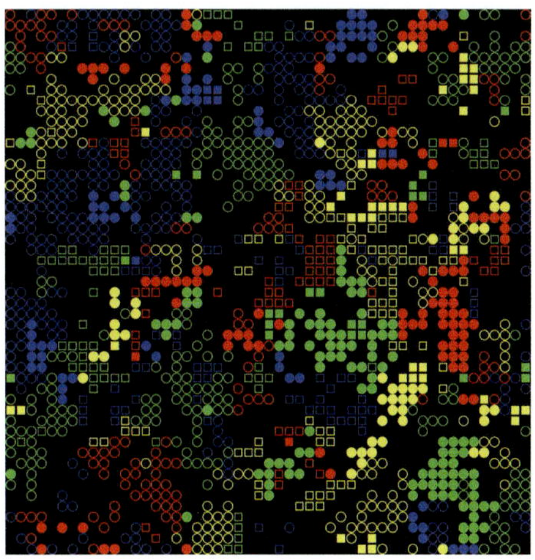

FIGURE 4.10: Ethnocentrism model

In the *Ethnocentrism* model (Hammond and Axelrod, 2006), each agent has one of four strategies: whether or not it assists (cooperates) with agents of the same colour as itself, and whether or not it assists with agents of a different colour. Each agent also has one of four colours (unrelated to the strategy). Assistance takes the form of giving up some of its own chance to reproduce to increase its neighbour's chance to reproduce. When an agent reproduces, the new agent has the same colour and strategy as its parent, so strategies that confer an evolutionary advantage come to dominate the population (Figure 4.10).

The final library model to explore is the *Traffic 2 Lanes* model (in the Social Science section). This model returns to our initial presentation of complexity, demonstrating how a traffic jam can emerge from drivers with different desired speeds reacting to the variations of speed of the driver in front of them.

4.3 Design of Tutorial Model

Now that you are more comfortable with agent-centric thinking and the way in which behaviour mechanisms at the individual level have consequences for complex system behaviour, we are ready to think about the design of the tutorial model. The tutorial model is obviously predefined, a model of protective behaviour in response to an epidemic. But we will go through the same design steps and questions that you would need to follow for your own model.

This will include broadly designing the model, identifying the key entities, interactions, and behaviours that are to be included in the code. Such design is an important step in any modelling process and the design is typically refined as the model and the modeller's thinking evolve. The design will be revisited throughout the tutorial to contextualise the coding of each element.

The first consideration is what to model. In most situations, the research question is decided first and then agent-based modelling would be chosen as the methodology because it is the most appropriate to respond to that particular question. Many questions concern whether an individual level mechanism is a plausible explanation for an observed system behaviour (such as the *Segregation* and *Ethnocentrism* described in Sections 1.5.1 and 4.2, respectively). Agent-based models are also useful for 'what if' explorations (or *justified stories*, see Section 15.4), for example to gain insight into the potential consequences of policy options. Additional types of questions are discussed in Section 2.5.2.

The question for this tutorial is of the 'what if' type: How is epidemic spread affected by protective behaviour? Potential scenarios could examine the effect of increasing the attitude of all individuals by some amount to inform a communications campaign, or identify whether there is a minimum efficacy for a protective measure to have any significant reduction in epidemic impact.

Having determined the question, it must be refined into a conceptual design for the model. There are three key elements: process, entities, and interactions. The conceptual design would typically include operationalisation of these elements, but these aspects will be partly deferred until required during model construction.

Operationalisation

Operationalisation is the process of specifying the detailed model rules that are to be translated into code. This includes how each relevant feature is measured within the model, how the relevant features that influence any action are combined, what is known to each agent, and each possible situation and its consequences.

As an agent-based model is a simulation, the focus of the design is the process or processes to be modelled. There are two processes in the tutorial model, adoption (and abandonment) of protective behaviour, and the epidemic (which includes both infecting new people and disease state changes for infected people). In a real model, the design for each process would be informed with existing subject matter knowledge, advice from subject matter experts (including disciplinary and local stakeholders), experiment, empirical data, and published literature.

Thinking about your model: conceptual design

1. What process (or processes) is to be simulated?
2. How to identify the relevant features and relationships for each process? Possibilities include: existing subject matter knowledge, advice, experiment, observational data, literature.
3. What agents are required to implement the process?
4. What characteristics of the agents are relevant to the process? This includes both characteristics that affect actions and status indicators.
5. What form does the environment take: abstract spatial, realistic spatial, social network...?
6. What environmental features influence the process or are affected by the process?
7. How do the agents and environment interact? This includes indirect interaction such as perception as well as input to actions and consequences of actions.
8. How to operationalise the process(es)?

Consider first the protective behaviour process. This process is a series of decisions by individuals or people, the main agent in the model. For the purposes of this tutorial, we assume that there are only three influences on a person's behaviour: their attitude or belief that the protective behaviour is useful, the behaviour of people around them, and the extent to which they feel threatened by the epidemic. This brings in one personal characteristic, one for social norms, and one that responds to the other process being modelled - the epidemic.

Attitude is to be operationalised as a value from 0 to 1, norms as the proportion of nearby people who have adopted protective behaviour, and threat as a function of recent nearby new infections. In addition, people communicate with their friends about threat in their location. If the weighted sum of attitude, norms and threat is above a threshold value, then the person will adopt (or maintain) the protective behaviour, and will abandon it once that weighted sum falls below the threshold. Further details of the protective behaviour operationalisation are provided in Chapter 7, where this process is coded.

The second process is that of epidemic spread. For influenza, the standard mathematical operationalisation is the SIR model introduced for the *Virus on a Network* model (see Section 4.2.2). Although influenza is spread from person to person, it is not necessary to individually model infectious people coming into contact with susceptible people. The virus survives on surfaces and in the air for several hours so it is reasonable to instead treat the epidemic at the population level, and infectivity as a characteristic of the spatial environment.

Epidemic spread is to be operationalised with differential equations that describe the change in the proportion of the population in each of three states: susceptible (never infected), infected, and recovered. The equations will be spatially explicit, so that the infected population can only infect others who are located nearby. Further details of the epidemic operationalisation are provided in Chapter 6, where this process is coded.

The two processes interact with each other. If the epidemic is active nearby, with a large number of new infections, that increases a person's likelihood to adopt protective behaviour via the threat component in the behaviour operationalisation. As the proportion of people who have adopted protective behaviour increases, epidemic transmission reduces, with the reduction also depending on the efficacy of the protective behaviour. The influence flows of the conceptual design are summarised in Figure 4.11.

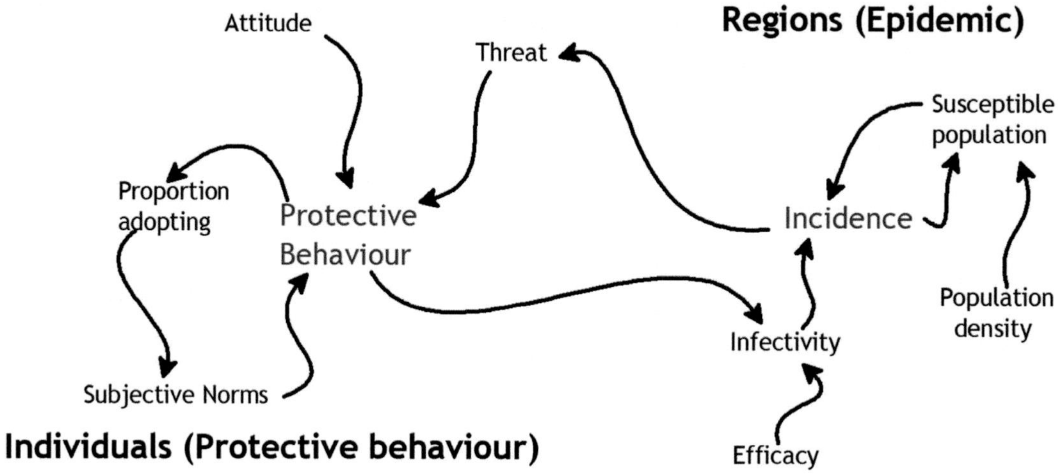

FIGURE 4.11: The broad design of the tutorial model. Individuals adopt protective behaviour based on some combination of their own attitude, the extent to which they observe other individuals adopting protective behaviour, and their perceived threat. Protective behaviour affects the transmission potential of the epidemic, which in turn affects the actual spread and hence the perceived threat.

The key elements of the design are:

- Agents: people
- Environment: spatial for epidemic, social for information about threat
- Resources: none
- Interactions: agent/agent - norms, environment/agent - threat, agent/environment - reduction in infectivity
- Attributes: agent - attitude, environment - population in epidemic states

The conceptual design is only one of the steps in designing a model (Badham et al., 2019). It is also important to design the interface, identifying the parameters that the user may wish to adjust and creating easily interpreted results and visualisations. The design may also consider the data that will be available for calibration (see Chapter 14) to ensure that there is sufficient detail to set appropriate values for the operationalised processes.

4.4 Structure of the Tutorial

The tutorial is constructed so as to teach three aspects of agent-based modelling simultaneously. **(1) Agent-centric thinking** is interpreting the world to be simulated as individual agents taking actions that implement some process. The particular computer programming language we use to build the model is **(2) NetLogo**. However, programming is not simply a matter of learning what each defined keyword in a language does, but **(3) good programming practices** are required to use that language effectively.

The remainder of the tutorial starts from an empty model. You will enter sections of code into the Code tab and create controls on the Interface tab (like the buttons, sliders, and plots in the library models). The complete instructions for building the model are presented in snippets, each of which contains both instructions (upper section) and text to be typed into the Code tab (lower section). Simply typing in the text will not create a working model. The controls on the Interface tab are automatically linked to the code and instructions for creating those controls must also be completed.

The snippets are organised so that the model is constructed incrementally. At the end of each group of snippets, there will be an instruction to run the model. Trying to run the model at other points in the tutorial will not break anything, but may generate errors.

Incremental programming

Add your code as gradually as possible, the smallest change that is complete. Then test it before moving to the next piece.

It is important that you actually *do* the tutorial. This is more than just following the instructions, typing in the programming code, and running the model. It is also important to think about how the code works. The particular commands are explained in the text and we describe what each section of code does, but you need to be sure that you understand how that code achieves its objective.

Key concepts are introduced immediately before you are expected to enter code to implement that concept. Don't worry if the concept does not immediately make sense; there will be repetition. However, you should spend some time thinking about how the code relates to the concept that has just been introduced.

In addition, any keywords that you are expected to learn are highlighted in a separate box. Early in the tutorial, only some of the keywords will be highlighted and the others will be deferred to later. By the end of the tutorial, all new keywords will be highlighted. Whenever you see a highlighted keyword, you should try to understand it, but also look it up in the NetLogo Dictionary (see the Help menu, or the NetLogo website).

4.5 Tutorial Progress Check

That completes the preliminaries, you are now ready to start constructing the model. You should be able to navigate an existing NetLogo model, switching between the interface where you use the model and the code tab that contains the program. You should also be able to use a NetLogo model to explore the behaviour of a complex system by running the model with different inputs, and understand why an agent-centric perspective of a process provides insight into that behaviour.

You should also have a general sense of the structure of the tutorial. There will be a short discussion of some agent-modelling or programming concept, then a set of instructions (including code to enter) that demonstrate that concept. At various points in the tutorial you will be instructed to run the model and also to reflect on the most important concepts.

Finally, you should have a broad picture of the model that is to be built in the tutorial. In particular, the model will contain individuals making decisions about protective behaviour based on their own attitude, the behaviour of other individuals around them, and their perception of the epidemic threat. Simultaneously, an epidemic will be simulated in the environment. These two processes influence each other and their interaction generates the shape of the epidemic spread through feedback: greater spread leads to higher perception of threat leads to greater adoption of protective behaviour leads to slower or even no spread.

5

Model 1: Model Entities

The first version of the model establishes the world and the agents that populate the world. While building it, you will also be introduced to some basic NetLogo concepts and take the first steps in good programming practices. Processes are not yet being modelled, there is no behaviour or simulation as there is no progression of time.

5.1 The NetLogo World

We will be starting with a completely empty model. There are two steps: creating a file, and defining the world. These are described in detail in the following, and are summarised at Snippet 0 for completeness.

5.1.1 Create a new model

Open NetLogo. If you already have it open from the previous activity of running library models, create a new model ('New' in the 'File' menu). You should have an empty model as in Figure 4.1.

Save the model ('Save' in the 'File' menu) in whatever file location you prefer. The name is arbitrary, but an appropriate name is *epidemic*, or *epidemic v1* if you want to save multiple versions at different points in the tutorial. If you do choose to save multiple versions, simply use the 'Save As ...' option in the 'File' menu when you want to start a new filename.

The file will be created with the extension *nlogox* automatically. That file extension lets the computer know that it is a NetLogo file. It also means that you can in future open NetLogo with the model already loaded by clicking appropriately on this filename in the file manager.

5.1.2 A world of patches

The NetLogo world (the area numbered (7) in Figure 4.1) comprises a grid of patches. Patches are one of the fundamental objects in NetLogo. They store information about the spatial environment and are also used for visualisation.

Each patch has a pair of co-ordinates, `pxcor` and `pycor`. The default geometry has 1089 patches in the 33×33 layout, with patch 0 0 at the centre of the world and both `pxcor` and `pycor` ranging from -16 to $+16$. Furthermore, the world is wrapped; for example an agent that moves off the right of the world reappears at the left.

pxcor, pycor

`pxcor` and `pycor` are the horizontal and vertical co-ordinates respectively of a patch.

For the tutorial model, increase the number of patches to 51×51 and turn off the wrapping. We also want to reduce the size of each patch so that the additional patches don't make the world too large for the screen. These changes are all made from the settings dialogue box.

Press the 'Settings' button (top right of screen) to open the dialogue box that controls the World. The dialogue box is displayed at Figure 5.1, with the settings to be changed highlighted. Change the values of 'max-pxcor' and 'max-pycor' to 25, and the patch size to 8. Uncheck both the horizontal and vertical wrap checkboxes. Once the settings are correct, press the 'OK' button to apply the settings and close the dialogue box.

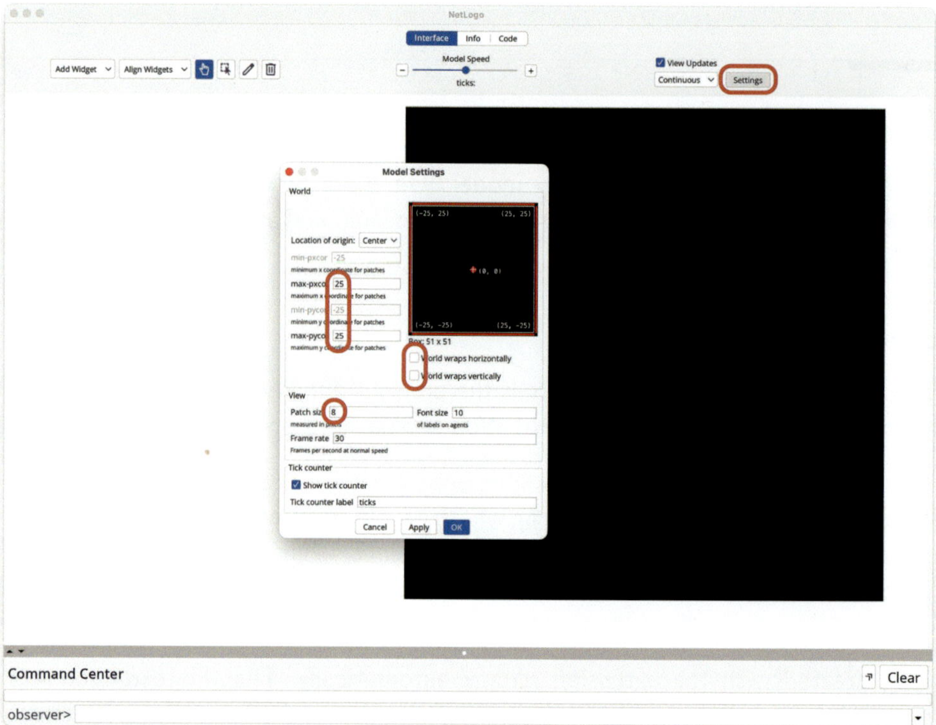

FIGURE 5.1: Screenshot of an empty NetLogo model with the Settings dialogue box open. Click on the 'Settings' button at the top right of the screen to open the dialogue box. The highlighted settings should be changed to match the values displayed.

The World and other interface components can be positioned anywhere within the NetLogo interface. A pop-up context menu is available (typically accessed by right-click on Windows systems, Ctrl-click on Mac OS) that has choices for Edit, Select, and other options. Note that the Edit option is an alternative way of accessing the Settings dialogue box because these are the settings to edit for the World.

First, make the Interface larger by dragging its edges or corners (with the mouse). Second, choose the Select option from the context menu. A box will surround the World with eight small squares.

That box indicates that the World is currently selected. Dragging the box will reposition the World and dragging the small squares will resize it. Drag the World to the right hand side of the now larger Interface.

Snippet 0: Create new model

1. Open NetLogo (if closed) or create a new model (if open).
2. Amend world settings as shown in Figure 5.1.
3. Adjust the window size and World location.
4. Save the (empty) model.

5.2 Fundamental Coding Concepts

At this point, we are ready to start entering text into the Code tab. The name *NetLogo* is used for both the software that runs models like those already opened, and also the programming language in which the model is written. This means that NetLogo is more than the set of keywords that provide instructions about how the model works (the code), but also the architecture that allows moving a slider on the Interface tab to change the way in which the model interprets those instructions. Both the code and the interface controls will be introduced incrementally.

5.2.1 Procedures

The main NetLogo code is comprised of procedures: blocks of code that perform a task. Each procedure is delimited by the keywords `to` and `end`, and the name of the procedure immediately follows the keyword `to`. You, as the model builder, can name the procedure whatever you like, but good practice is to use a name that is reasonably short but also descriptive. For example, the names of procedures already seen include 'eat-grass' (Figure 4.6) and 'spread-virus' (Figure 4.7).

to <procedure-name> <instructions> end

A NetLogo procedure is a block of instructions to the computer. The keywords `to` and `end` mark the beginning and end of the block, respectively. The first word following the keyword `to` is the name of the procedure.

A second type of procedure is delimited by the keywords `to-report` and `end`. These are discussed later in the tutorial (see Section 7.3).

Many procedures call other procedures. That is, the block of code for one procedure includes the name of another procedure (referred to as 'nested'). NetLogo carries out the instructions line by line, starting at the top of the procedure. When it gets to a call to a different procedure, execution moves to the start of that other procedure, running from the top to the bottom. Control is returned to the original procedure when the `end` keyword is reached, and any remaining instructions are completed. NetLogo programs can have several levels of procedure, with one procedure calling several others, which each call further procedures and so on. For example, in the *Virus on a Network* model, the 'go' procedure called the 'spread-virus' procedure, which (with some probability) called the 'become-infected' procedure (as seen in Figure 4.7). Eventually, the procedure that is deepest in the nested calls reaches its `end` statement and the execution steps back up a level and continues until either a call to another procedure or the `end` of that higher procedure.

5.2.2 Comments

All programming languages have a way to leave comments in the code. These are for a human reader to help them interpret what the code is intended to do (Knuth, 1984). The instruction to the computer is to simply ignore them. In NetLogo, a comment is indicated by a semicolon (;). The semicolon and anything after it on the same line is skipped when the computer executes the program.

comment (;)

The computer ignores anything appearing between a semicolon (;) and the end of that line. It is used to indicate a comment from the programmer to any humans reading the code.

Comments are useful to explain what a section of code does but, also, why that section of code is needed. For example, the comment in the *Rabbits Grass Weeds* 'move' procedure (Figure 4.5) states that 'moving takes some energy', which explains the purpose of the following line of code where energy is reduced. The line of code is straightforward to interpret, so it would be uninformative to have a comment that simply says 'reduce energy' and does not explain why.

In this model, we also use comments as headings for groups of procedures. The comment used for the group heading is written over multiple lines to make it easily visible when scrolling through the model. Procedures can be organised however is convenient for the modeller; the order does not matter to the computer. The drop-down list of procedures is ordered in the same way that the procedures are placed in the code. Organising your code into groups of related procedures makes it easier to quickly find the one you need.

5.3 Start and Continue

The current state of the real world is a snapshot in time. Focusing on the process of interest, the current situation incorporates what has already occurred in the past. In some sense, time can be broadly broken into past, present (the current situation), and future.

A simulation, however, must start at some specific point in time so there must be an initial state. The past is represented in the simulation only by its effect on the initialisation. Of course, the past from our current perspective can be simulated by setting the initial point of time even further in the past, but that simply moves the problem.

All simulations differentiate between initialisation (the starting state) and the rules that step through the process being simulated (see Figure 5.2). In NetLogo, it is conventional to construct this differentiation as two main procedures.

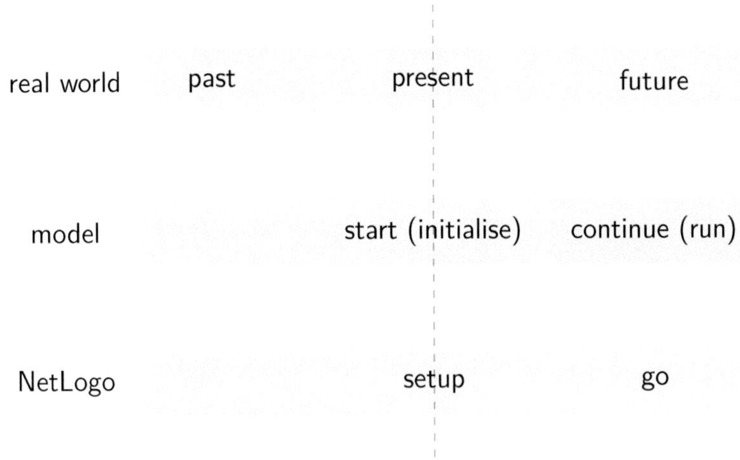

FIGURE 5.2: A model must be initialised to describe the state of the world at some arbitrary start point, which also captures the outcomes of actions taken in the past. It is then run to simulate the future. In NetLogo, this separation is typically constructed in a *setup* procedure for the initialisation, and a *go* procedure for the actions that occur during the model run.

The *setup* procedure contains all the code necessary to initialise the model. This would typically include creating all the agents and environment, and setting values for attributes of the agents and features of the environment. For example, in the *Rabbits Grass Weeds* model, the *setup* procedure created rabbits and grass, and assigned starting energy values for each rabbit.

The *go* procedure contains all the code necessary to implement the process. This is the code that contains the interactions, decisions and actions that describe the model mechanisms and change the agent attributes or environmental features. For example, in the *Rabbits Grass Weeds* model, the *go* procedure had the rabbits move, eat, reproduce, and die.

5.4 Create Model Control Buttons

Almost all NetLogo models have two main control buttons; by convention these are named 'setup' and 'go'. The setup button is used to generate the starting point of the simulation, and the go button is used to run the simulation by stepping through time.

Buttons are just one of the many widgets available. A widget is any component of the interface, typically used to control the model or report information from the model. In the library models, you have used buttons to start the model, sliders to enter values for model settings, and seen plots of values generated by the model. Buttons, sliders, and plots are all types of widgets. Technically, the World display is also a widget.

To add a button, use the 'Add Widget' drop-down box (circled in Figure 5.3) to choose a Button. Move around the interface and click to position the button. Alternatively, you can add a widget by opening the pop-up context menu in the interface (right-click for Windows; Ctrl-click for Mac OS) and selecting 'Button'. Either method will open a dialogue box (Figure 5.4) to request the settings for the button you wish to add.

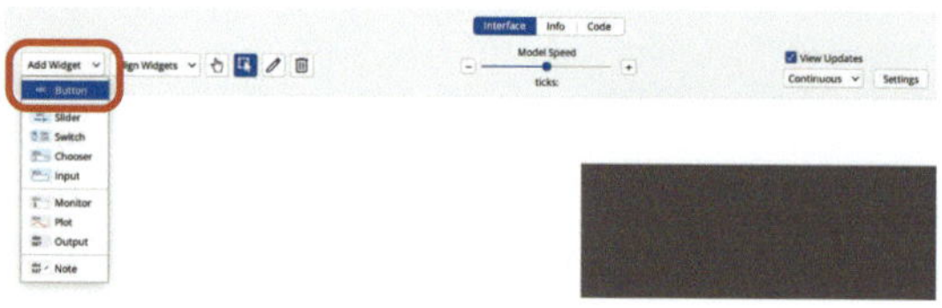

FIGURE 5.3: The menu item to open the dialogue box to create a button on the Interface. The drop-down box (circled) is used to select the type of control widget to add (in this case, button). Then point to anywhere on the Interface and click to position the button.

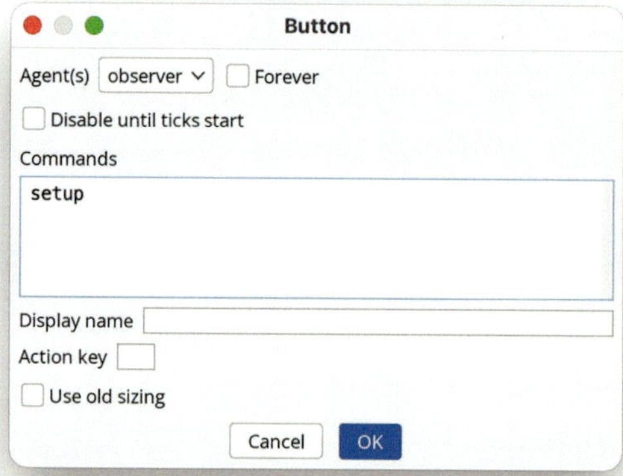

FIGURE 5.4: Dialogue box for setup button.

First create the setup button. As in Figure 5.4, enter the word 'setup' in the *Commands* box, and also in the *Display name* box. Press *OK* and a button will be created labelled 'setup'. The display name is the text to place on the button. The command box can contain any valid NetLogo commands. In this case, the button will call a procedure named 'setup'. However, that procedure does not exist so the button will be red to indicate an interface error.

Create a second button. For this one, the command is 'go'. In a typical NetLogo model, the 'go' button would call a 'go' procedure and be labelled with the text 'go'. However, to demonstrate that this consistency is not required, set up the button to call a procedure named 'go' (command box) but label it with 'do it!' (display name box). For this button, check the 'Forever' checkbox. Checking 'Forever' adds a circular pair of arrows icon to the button but will not actually have an effect until time is introduced to the model. Press *OK* to complete the second button.

The next step is to enter the code in the bottom part of Snippet 1a. Switch to the *Code* tab and simply type it in. The code contains the two procedures called by the buttons, but neither procedure has any content. Instead, each simply has a comment (make sure you have a semicolon at the beginning of each line). In addition, there is a heading comment stating that these are the main control procedures. The box style frame and upper case is intended to make the heading more visible as you scroll through the code.

As soon as you start typing in the *Code* tab, the *Check* button (top left, circled in Figure 5.5) becomes active. Once you have completed typing in the code, click on that *Check* button. This will check the code for errors such as mismatched brackets and words that NetLogo doesn't know what to do with. These are typically typographical errors such as misspelling a keyword or procedure name. If there is such an error, a red bar will be placed at the top of the screen together with an error message (see Figure 5.5). In addition, the *Code* tab label will be red to indicate there are unresolved errors. Unfortunately, there are many errors of other types (such as logic errors) that NetLogo does not identify automatically.

If you do have errors, make the appropriate changes to the code and *Check* the code again. Once there are no errors (and the *Check* button returns to inactive), save the model. Look at the interface; the button labels should be blue (instead of the previous red). Now that the procedures exist, NetLogo is able to interpret the button command to call the procedure.

Snippet 1a: Buttons

1. Create a button labelled 'setup' to call the procedure named 'setup'.
2. Create a button labelled 'do it!' to call the procedure named 'go', with Forever.
3. Enter the following code to create the 'setup' and 'go' procedures.
4. Check the code and Save the model.

```
;-----------------------------------------------------
; MAIN CONTROL PROCEDURES
;-----------------------------------------------------

to setup
  ; initialisation procedures go here
end

to go
  ; model step procedures go here
end
```

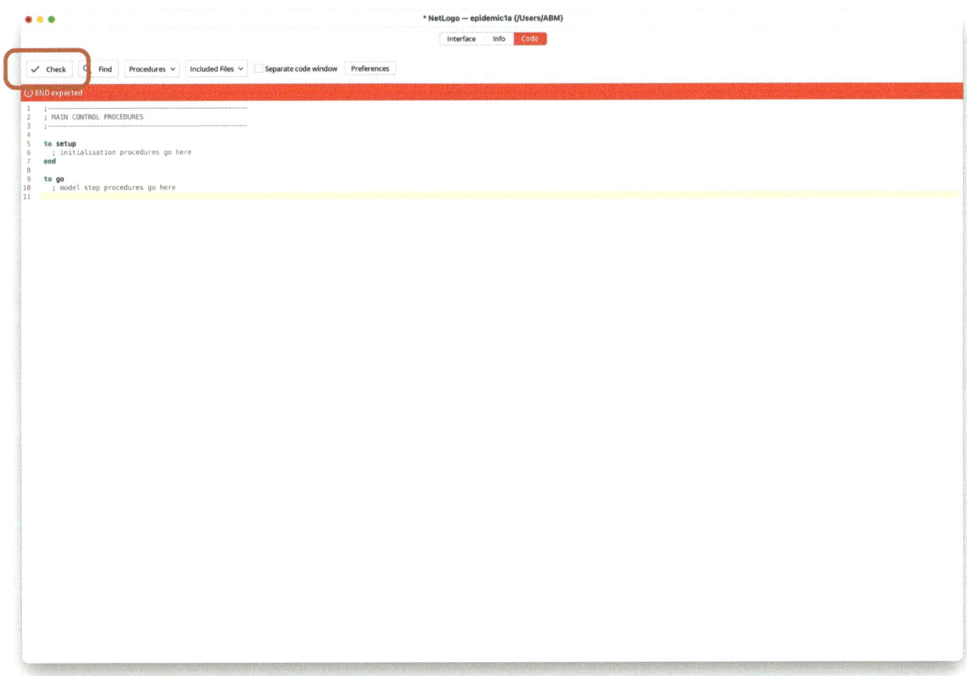

FIGURE 5.5: Screenshot of a typical error message presented after clicking on the *Check* button (circled). In this case, the word **end** is missing to end the 'go' procedure, so the error message is describing that error and the yellow shading indicates the position where the error was found.

You can now run the model. Go to the interface and press the *setup* and *go* buttons. Nothing will happen, but NetLogo should not give an error, either.

Buttons can be moved in the same way that the World can be moved. Access the context menu, press 'Select' and use the frame to drag or resize the button. Selecting a widget in this way will also activate the widget selection tool and you can select multiple widgets and use the 'Align Widgets' drop-down to make the interface neater. The finger pointing icon exits from widget selection. Once the widgets are in their appropriate places (see Figure 5.6), model version 1a is complete. Save the model (Save item in the File menu).

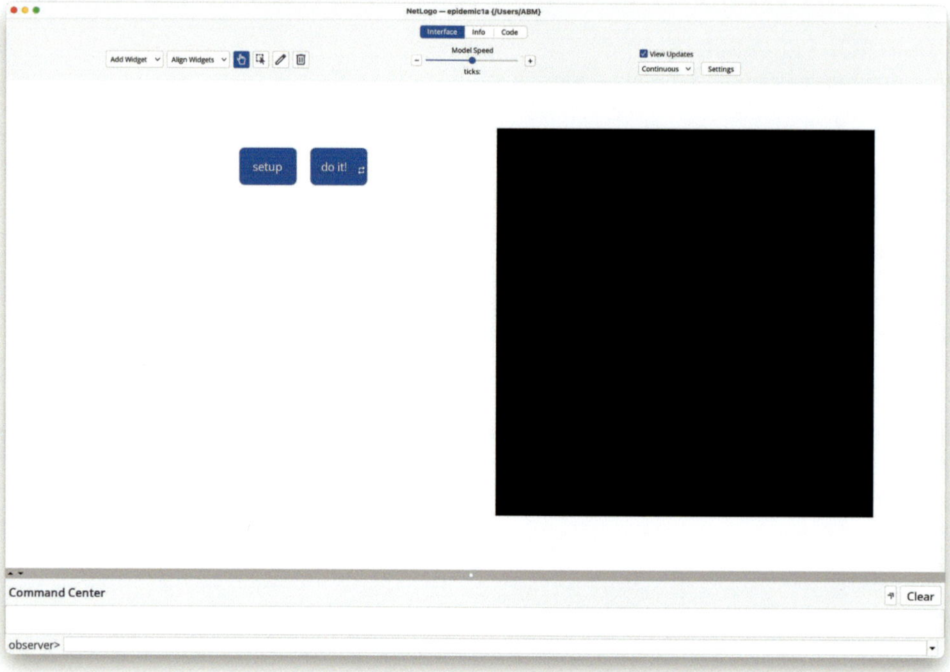

FIGURE 5.6: Screenshot of the interface after completing version 1a of the model.

5.5 Variables

A variable is simply a way of storing information. You can think of it as a box with a labelled lid. The actual contents of the box can change, and the box may even be empty, but the label stays the same. Variables provide a convenient way of keeping track of information that the computer needs to know, such as the energy of a rabbit or the speed at which grass grows in the *Rabbits Grass Weeds* library model, and then referring to that information as required in the code.

The value of the variable (or contents of the box) can be a number (eg 10), string (eg "high"), boolean logic value (`true` or `false`), agentset (see Section 7.4), or list (see Section 12.2). There is no need to specify the type of contents for the variable.[1]

[1]Computer scientists would refer to this as dynamic typing. Once the first value is stored in the variable, then NetLogo knows what type of value it is and will report an error if you try to do something that doesn't make sense, such as asking to add a number to a string. The error will not be detected until you first try to run the model. The programmer needs to remember what they are using the variable to store.

Variables

A variable is a way to link a meaningful name to a piece of information that can be stored, changed, and retrieved. There are three types of variables in NetLogo: global variables contain information available to all model entities, agent variables store information about individual agents (including patches or links), and local variables are temporary storage.

5.5.1 Variable names

Variable names are for the convenience of the programmer (and model user), and should therefore describe the information being stored. On the other hand, variable names should be fairly short so that they can be accurately typed when writing the code, and also not interfere with reading the code. Only certain characters are permitted in names: letters (a-z, A-Z), numbers (0-9), and some special characters (including ? # - _ %).[2] Other special characters (including @) will create an error if included in a variable name. The name can start with any of the permitted characters except for a number.

The most common convention in NetLogo variable (and procedure) names is to use all lower case and to join words with a hyphen (-) for more descriptive names. This leads to some potential confusion between variable names and procedure names but, in practice, variable names tend to be single words and procedure names tend to be several words so the confusion rarely arises. This is the convention used in the NetLogo Models Library and in this tutorial.

Some other naming conventions for joining words are to use an underscore (_) or camel case (upper case at the start of each word, for example 'ExampleVariableName'). The convention you adopt is not important as long as you adopt a convention; consistency reduces opportunities for syntax errors.

Apart from connecting words, symbols are not generally used in NetLogo names, with three important exceptions. The first is that a question mark (?) is always used as the last character for a variable that is only true or false, and never used in names otherwise. The second exception is that some models use a special character to start variable names within procedures that take arguments. In this tutorial, the hash character (#) is used for that purpose (see Section 7.3), but the underscore (_) is also common. Finally, variables that contain percentage amounts often have the percent character (%) at the end.

5.5.2 Accessing variables

NetLogo must be informed about a variable's existence before it is actually used. This allows NetLogo to allocate some storage and link the name to that storage. Declaring a variable's existence is done differently for each of the variable types.

Global variables are declared at the top of the code with the `globals` primitive followed by a list of variable names. In addition, widgets such as sliders that take values from the model user automatically create a global variable of the name given to the widget. Variables that store information about an agent or link are created when the agent or link is created. However,

[2] Technically, mathematical operators are also available, but these should be avoided as it is too easy to confuse manipulating multiple variables with referring a single variable (for example, income + capital adds two variable values together, and income+capital is a variable name).

NetLogo needs to know what variables to create and they are also declared at the top of the code, but with `turtles-own` and similar statements. Similarly, a `patches-own` statement is used to name all the variables associated with each patch.

Local variables are temporary storage that is forgotten once the relevant piece of code has finished running. These are created with a `let` statement, which both declares that the variable exists with the specified name, and stores the results from some piece of code into that variable.

The contents of the variable that is known to a model entity (like a patch or turtle) are accessed simply by using the name of the variable. This includes global variables and the agent's own variables. In addition, turtles are aware of the variables owned by the patch where they are located. However, if an entity wants to use information that it does not automatically know, then the code must also specify the entity that owns the relevant variable (using the `of` primitive).

5.6 Describing the Physical Environment

A key factor in the spread of an influenza-like epidemic is population density. As the epidemic is to be simulated by the patches, representing the physical environment, the patches will need a variable for their population density. However, all patches are of the same size, so we will instead use population count because that makes the transmission equations easier conceptually.

We will also introduce a global variable for the maximum population of any patch and show its contents on the interface. This allows a soundness check, which is a quick assessment of whether results make sense. For example, if we write code that is intended to assign population values in the thousands, but the maximum population shows millions, then there is an error in the assignment method. Similarly, if the maximum population shows 0, then population values are not being assigned. Such soundness checks are essential while building any model to make sure that the model being built is consistent with its design.

Create a variable to store the maximum population value (Snippet 1b-1). This snippet introduces the `globals` keyword, which is used to list the global variables. Note the short descriptive variable name, and that a comment has been used to provide a more detailed description.

Snippet 1b-1: Declare global variables

1. Enter the following code, starting at the first line at the top of the *Code* tab.
2. Check the code and Save the model.

```
globals
[ max-popn      ; maximum population of a patch (for colour scale)
]
```

globals

The `globals` keyword is used to identify all the global variables except those that are controlled by the interface widgets that provide information to the model (sliders, switches, choosers, and inputs). The variable names are listed between square brackets.

This code only instructs NetLogo to create the variable, its value is calculated later (in Snippet 1b-3). First, we need to create a monitor to display its value. A *monitor* is an interface widget to show the contents of a single variable. It is created in a very similar way as a button. Go to the *Interface* tab and add a monitor (change the widget drop-down to 'Monitor' and click on 'Add', or access the context menu).

The dialogue box (see Figure 5.7) has a box for NetLogo code and the Display name, which is the same as the button widget. It also has settings for how the results are displayed, such as decimal places (change to 0). As the contents of a variable are accessed simply by using the name, the variable name is the only code required in the Reporter box.

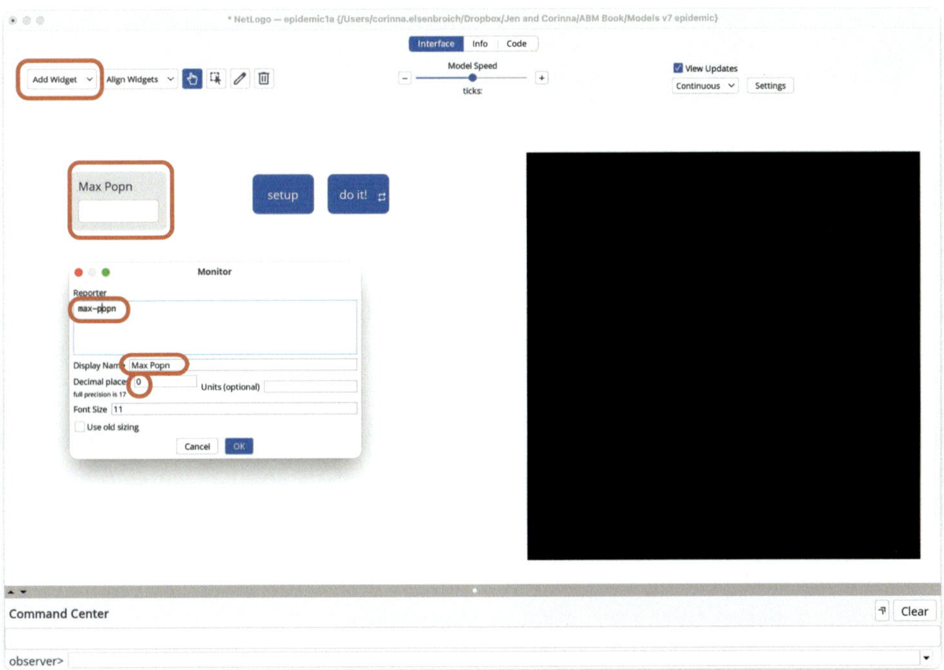

FIGURE 5.7: Screenshot with dialogue box for maximum population monitor. Note that the title of the monitor matches the text in the 'Display Name' box.

At this point, the monitor should display 0. A NetLogo variable, once created, has a value of 0 until it is `set` to some other value. The global variable named max-popn is created by the `globals` statement and assigned the value of 0 by default.

However, the variable does not yet have any meaning. Patches need a variable to store their population and then population values assigned for each patch. After that, the maximum can be calculated and assigned to max-popn.

For *patches* (and *turtles* and *links*), the variables available are declared with the `patches-own` keyword (and similarly for the other entities). Each patch has its own copy of the variable with potentially different values. First, we must create a variable to store the population value for each patch, to be named 'popn' (see Snippet 1b-2).

Snippet 1b-2: Declare patch attributes

1. Enter the following code below the globals declaration but above the heading for the main procedures.
2. Check the code and Save the model.

```
; patches have epidemic information
patches-own
[ popn                ; population of patch
]
```

patches-own

The `patches-own` keyword is used to identify all the variables (listed between the square brackets) that are available to patches to describe attributes of the spatial environment. Each patch has its own value for each of the variables named.

Having made the variable available, the next task is to assign a suitable value of population to each patch. In the real world, population density varies greatly. Typically, most people will live in one of a small number of cities. For the model, three patches will be assigned large population values, with small values for all other patches.

Snippet 1b-3 creates a new procedure (called setup-patches) that uses the `random` primitive to generate random numbers and assign them as population values. It also calculates the maximum of these and changes the value of the max-popn variable to that maximum. The new procedure is called by the setup procedure simply by including the procedure name in the code for the setup procedure. This means that clicking on the setup button starts the setup procedure, and the setup procedure in turn diverts to the setup-patches procedure.

random

Generates an integer between 0 and n-1 where n is the number immediately following the word `random`. Each integer in the range is equally likely to be generated.

In setup-patches, `random 10000` returns a randomly generated integer from 0 to 9999 and that number is added to 2000 to assign a population count (with the `set` primitive) to each patch. Three of the patches are randomly chosen for a higher population count, representing cities. For these, `random 3` is used to generate a random number from the set {0, 1, 2}, which is then added to 3 and multiplied by 1 million. So the city patches each have a population value of 3 million, 4 million, or 5 million. Other new primitives are explained later in the tutorial.

Snippet 1b-3: Assign patch populations

1. Edit the 'setup' procedure to delete the comment and replace with the code below, which also calls a new procedure named 'setup-patches'.
2. Create the 'setup-patches' procedure (in a new section at the bottom of the code with the heading comment 'Implementation Procedures').
3. Check the code and Save the model.

Note: Shading is used for surrounding code that does not change.

```
to setup
  clear-all
  setup-patches
end

 to go
   ; model step procedures go here
 end

;--------------------------------------------------
; IMPLEMENTATION PROCEDURES
;--------------------------------------------------

to setup-patches
  ask patches [ set popn 2000 + random 10000 ]
  ask n-of 3 patches [ set popn 1000000 * (3 + random 3) ]
  set max-popn max [popn] of patches
end
```

Having added some code, the next step is a soundness check; does it work as expected? The Check box (with tick icon) only checks syntax. Logic errors in agent-based models can be difficult to diagnose, so it is important to add code gradually and verify it before adding any further code since the code is intended to implement the model design.

Soundness check

Whenever you add code to your model, do a soundness check. Test the code to see if it does what you expect it to do. If not, fix it before moving on. The *Check* button (top left in code tab) only checks syntax, not logic.

The most obvious check is to press the *setup* button and look at the number in the 'Max Popn' monitor. From the discussion of how the setup-patches procedure is intended to work, the value should be 3, 4, or 5 million. Because the maximum is randomly generated, repeated button pressing should also lead to some changes in the number returned.

Note that a common error is to create a new procedure, but forget to include the call to that procedure in one of the main procedures. In this case, the value of max-popn would never change from its default of 0. You can see this by commenting out the call to the setup-patches procedure in the setup procedure.

This check only confirms that appropriate values are assigned for the patches representing cities. What about the others? Checking can be made more rigorous by commenting out the line that generates the city populations (by placing a semicolon at the beginning of the line), the maximum population value generated for other patches will be displayed in the monitor. You should expect to see a value slightly below 12,000. Remember to delete the semicolon after checking.

At this point, the model looks like the screenshot at Figure 5.8. Once the maximum population is correctly calculated after clicking the *setup* button, model version 1b is complete. Save the model (Save item in the File menu).

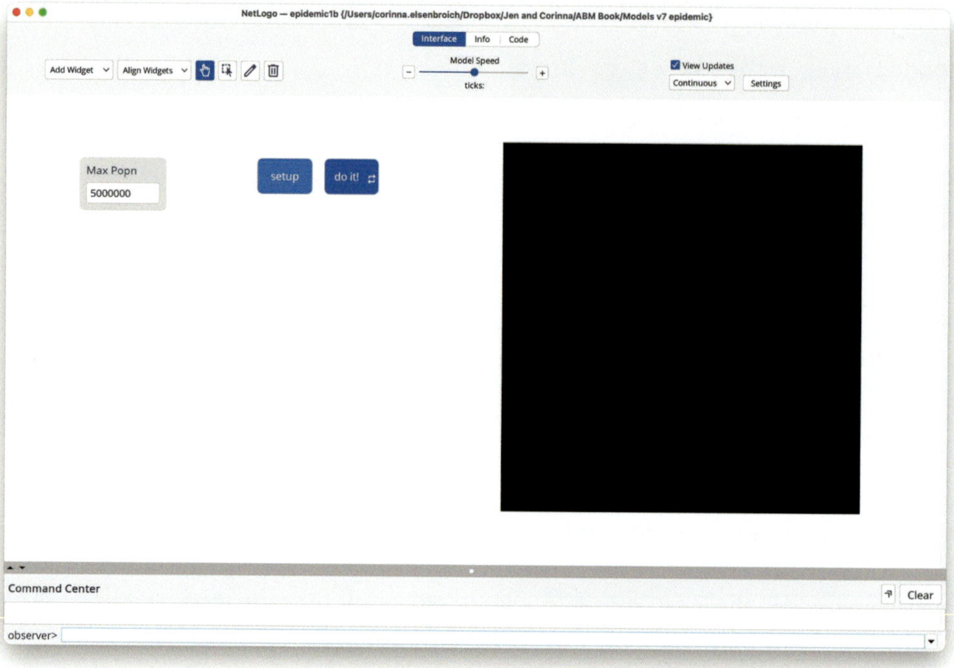

FIGURE 5.8: Screenshot of the interface after completing version 1b of the model and pressing the setup button.

5.7 Doing Mathematics

Consider equation 5.1:

$$x = 1 + 2 \times 3 - (4 + 5) \div 6^2 \tag{5.1}$$

How would you calculate the value of x? Hopefully not by starting at the left and working your way along (with $1+2=3$ then $3 \times 3 = 9$ then ...). The correct way is to start with the brackets, then do the 'power of' (called exponentiation), multiplication and division, then finish with the addition and subtraction. Breaking the calculation into small steps:

	step done:
$x = 1 + 2 \times 3 - (4 + 5) \div 6^2$	
$= 1 + 2 \times 3 - 9 \div 6^2$	$4 + 5 = 9$
$= 1 + 2 \times 3 - 9 \div 36$	$6^2 = 36$
$= 1 + 6 - 0.25$	$2 \times 3 = 6$ and $9 \div 36 = 0.25$
$= 6.75$	$1 + 6 - 0.25 = 6.75$

In mathematics, the convention for ordering of the calculation steps is referred to as precedence. Programming uses the same term, but the operators that must be considered are more than simply $+$, $\times$, and so on. Precedence also applies to logical operators (such as `and` or $>=$) and primitives that work on agentsets (see Section 7.4).

NetLogo, along with other programming languages, also uses specific symbols for these mathematical operators. Your computer keyboard has plus ($+$) and minus ($-$) symbols, but there isn't a key for the 'divided by' sign ($\div$), it would be confusing to use the letter 'x' for multiplication, and how would you indicate the superscript that is used for exponentiation when you have a pen and paper? The symbols used in NetLogo suggest the pen and paper symbols, but are not the same. Exponentiation uses the 'caret' symbol (^), also referred to as 'circumflex' or 'hat', pointing up. The asterisk ($*$) is used for multiplication, a sort of $\times$ with extra arms; and division is indicated with the forward slash (/), a standard alternative for $\div$ in equations.

Because NetLogo is very flexible with variable names, all mathematical operators must have a space immediately before and after. For example, 6 to the power of 2 would be written as `6 ^ 2`, not as `6^2`. The only exception to spacing is brackets, no space is required before or after either a normal round bracket or a square bracket.

Precedence

Precedence is the order in which pieces of an equation are calculated. Brackets are used to change precedence; anything in brackets is done first. The mathematical operators follow the standard mathematical precedence order of exponentiation (^), then multiplication ($*$) and division (/), then addition ($+$) and subtraction ($-$).

Putting the symbols and spacing rules together, and remembering that `set` is used to assign a value to a variable, equation 5.1 would be written as follows in NetLogo:

```
set x 1 + 2 * 3 - (4 + 5) / 6 ^ 2
```

NetLogo also offers a range of other mathematical functions. These are all implemented with primitives rather than symbols. For example, the square root of 4 would be written as `sqrt 4` and the logarithm can be found with `log` (base 10) or `ln` (base *e*). Also available are statistical functions for manipulating multiple values (eg `sum`, `mean`), trigonometric functions for geometry (eg `sin`), and logical operators (eg `and`) and comparisons (eg >=). The full list is available in the NetLogo Dictionary (Help menu).

Spaces in code

NetLogo is very sensitive to the presence of space. Every keyword, operator, procedure name, variable name, mathematical operator, or other code element must be separated with at least one space. The only exception is brackets; so any of (or) or [or] can be used without leaving a space. The number of spaces does not matter; any number of spaces is treated as a single separator.

5.8 Turtle Agents

Having established a world for our agents to occupy and perceive, we need to create agents to make decisions. In NetLogo, these decision-making agents are referred to as turtles.

The model developer is able to specify multiple types of turtles, referred to as breeds. This contrasts with patches, where there is only one type. Each breed of turtles has its own list of agent variables.

breed

Turtle agents can be of different types, referred to as a breed. To define a breed, the `breed` keyword is followed by the plural form of the breed name, then the singular form.

For the tutorial model, our agents are referred to as 'people' because it is people who decide whether to adopt or abandon protective behaviour. Those people need a variable to store their protective behaviour status. Snippet 1c-1 defines a `breed` of turtles called people (or person when referring to a specific turtle), then declares the variables available to people with `people-own`. Note that the variable is to store whether the person has adopted protective behaviour, so its value will be `true` or `false`, and therefore has a question mark at the end of its name.

turtles-own

The `turtles-own` keyword is used to identify all the variables (listed between the square brackets) that are available to turtles to describe their attributes. Different sets of attributes can be defined for different breeds of turtles, with the name of the breed used in the keyword instead.

The next step is to create people. Patches are part of the provided environment and automatically created simply by starting NetLogo. However, code is needed to inform NetLogo of the number and type of turtles to create.

There are three different primitives for creating turtles, depending on who is issuing the instruction (or whose perspective is being taken). The observer can `create-turtles`, patches can `sprout`,

Snippet 1c-1: Define agent attributes

1. Add the following code below the **globals** definition and above the **patches-own** statement.
2. Check the code and Save the model.

```
; people deciding about protective behaviour
breed [people person]
people-own
[ protect?              ; true / false for protected
]
```

and turtles can `hatch`. To create/sprout/hatch turtles of a particular breed, the (plural form) name of the breed is used instead of turtles. For example, the tutorial model will create turtles using **`sprout-people`** to instruct the patches to create turtle agents of the people breed (in Snippet 1c-2) shortly.

Why does NetLogo need three different ways of doing the same thing, creating turtles? Because they represent different process mechanisms, allowing better matching between the real world and simulated behaviour. In particular, `hatch` is reproduction, which occurs while the simulation is running, it is part of the behaviour. In contrast, `create-turtles` and `sprout` are typically used to create turtles during initialisation, to make them available for the simulation. The key difference between these two is that `sprout` is convenient for spatially distributed turtles.

Regardless of the primitive used to create the turtles, it is followed with a number, indicating the number of turtles to be created, then an optional block of code in square brackets (like [ ]) that each turtle is to run as it is created. This block of code is used to set values for the attributes. In addition, turtles created with `sprout` inherit their location from the patch doing the sprouting, and turtles created by `hatch` inherit variable values from its parent. In both cases, those values can be overwritten using a `set` statement in the block of code.

create-turtles, sprout, hatch

The primitives `create-turtles`, `sprout` and `hatch` are all used to create turtle agents. They differ by the type of entity responsible for creation: observer, patch, and turtle agent, respectively. Variable values such as location are inherited by the new turtle from its parent patch or turtle when created with `sprout` or `hatch`. For all three primitives, a `breed` can be specified for the turtle(s) to be created. A code block following the creation contains any code that is to be run immediately on the turtle's creation, typically containing initial values for variables.

Follow the instructions for Snippet 1c-2. For this model, multiple people are needed at each patch so that there is the opportunity for people to make different decisions and therefore contribute to different levels of protection. A larger number of people allows finer gradations in the proportion protected. This snippet uses the `sprout` primitive to create agents to control their location.

Snippet 1c-2: Create agents

1. Create the displayed 'make-people' procedure in the implementation procedures section of the code.
2. Add call to 'make-people' procedure in the 'setup' procedure.
3. Check the code and Save the model.

Note: Shading is used for surrounding code that does not change, and the ellipses (...) indicates that there is code that is not shown.

```
to setup
  clear-all
  setup-patches
  make-people
end

...

to make-people
  ask patches
  [ sprout-people (2 + floor ( popn / 10000 ))
    [ set protect? FALSE
    ]
  ]
end
```

The turtles being created are of the breed people. Comment out[3] the code created in Snippet 1c-1, then check syntax and NetLogo will highlight the sprout statement with an error. So the breed statement defines a type of turtle that is then available to be used.

Use a breed for turtles

Even if you only intend to have one breed of turtle agents, it is worth creating it with a **breed** statement rather than using the default. Firstly, this allows you to use a meaningful name for your agents, which makes the code more readable. Secondly, no recoding would be required if you later decide to add an extra breed of turtles.

The number of people to create is included in the code as a calculation rather than a specific number. Each patch has a value for the 'popn' variable and that value is used so that patches with a larger population also have more people. But an exact number of turtles must be created, NetLogo will report an error if the calculation results in fractional people. The floor primitive is a mathematical way of saying 'take the integer part'. For example, 19/10 is 1.9 and rounding to the closest integer would result in 2, whereas floor 19/10 is 1. in this case, the value of

[3] It is easy to comment or uncomment a block of code by highlighting the code (click and drag the mouse over the block) and then use the menu item Edit > Comment/Uncomment.

popn will be a few million in the city patches, which will consequently have a few hundred turtle agents, and a few thousand in the other patches, which will have either 2 or 3 agents.

If you accidentally have the call to the make-people procedure before the call to the setup-patches procedure (in the setup procedure), then the popn variable values will all be 0 when the number of turtles is calculated and two turtles will be created at each patch.

As each person is created, the new person runs the code in the square brackets to `set` the 'protect?' variable to `FALSE`. This variable is available because of the `people-own` statement in Snippet 1c-1. It is `set` to `FALSE` to indicate that the person has not adopted protective behaviour.

After pressing the setup button, the model looks like the screenshot at Figure 5.9. Each coloured icon in the World is one person. Once turtle agents are being created successfully, model version 1c is complete. Save the model (Save item in the File menu).

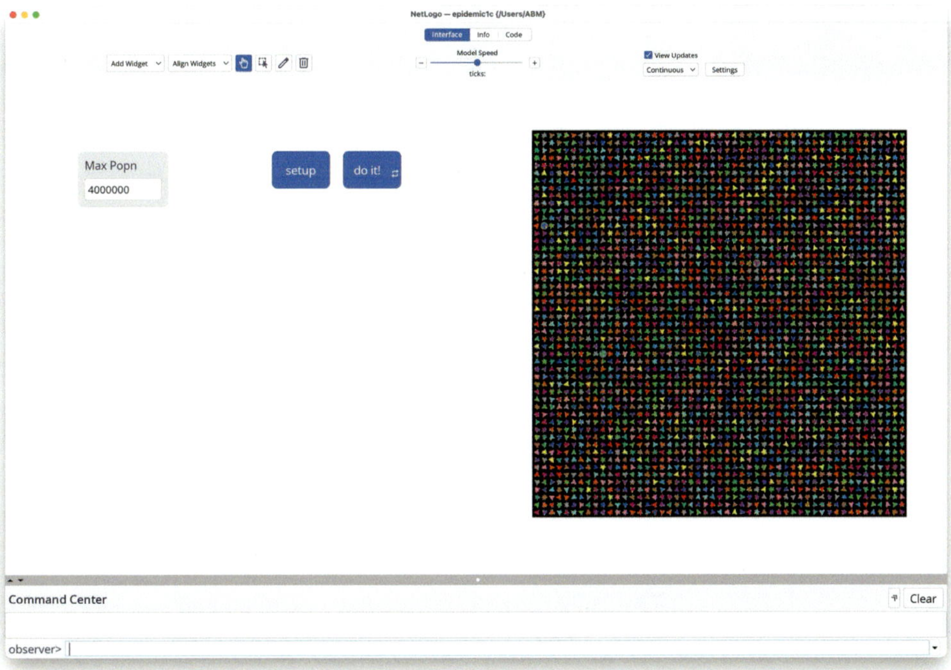

FIGURE 5.9: Screenshot of the interface after completing version 1c of the model.

5.9 Testing with Inspect Windows

What sort of soundness check could be used here? When you have learned more code, it would be easy to check the minimum and maximum number of people on any single patch. However, that option is not available yet. What about getting a closer look at the World instead?

Press the setup button so that the World is populated. Place the mouse pointer on one of the people icons and open the pop-up menu (with right-click in Windows and Ctrl-click in Mac OS). If you have the pointer on a turtle agent, the bottom of the menu will have some submenus, one for each person at the pointer. If the bottom line is 'inspect patch...', then move the pointer and try again. Open the submenu for one of the people and select the *inspect* option. This opens the inspect window (see Figure 5.10), which lets you look at individual agents in the model.

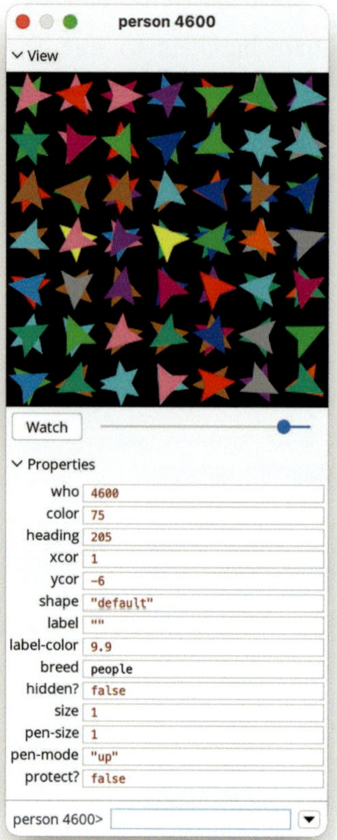

FIGURE 5.10: Inspect window for person.

The person you chose to inspect is identified at the top of the window - person 4600 in the example. The top section has a small area of the World, 7×7 patches at the default zoom level. Immediately below this display is a slider to control the zoom. This World section is 'live'; it is not just an image of the World. Any change in the World is also seen in the display and, for example, you can use the display to open further inspect windows.

The next section is a list of all the variables owned by the turtle (or person in this case) and their values. The first 13 variables are built-in variables that are automatically created for all turtles. They include icon controls (color, shape, size), the breed, and position (xcor, ycor). The variables created by the `turtles-own` keyword are then listed. The only variable so far created is named 'protect?' and can be seen as set to `false`. At the bottom of the window is a box where NetLogo commands can be entered.

You can also open inspect windows for patches (and links, once there are some in the model). Open the context menu again, and this time select the 'inspect patch...' option. The inspect window for a patch has the same structure as for a turtle: the title bar identifies the patch, with a section of the World immediately below, variables, and then a code box. For patches, there are only 5 built-in variables before moving to the model specific ones.

The only `patches-own` variable defined so far is named popn. Look at the popn value for the patch you are inspecting. From previous discussion, you should expect the value to be near 10000 or a few million. For cities, there will be several hundred people on the patch and therefore in the centre of the World display. But for most patches, there will be two or three people displayed in the centre of the World. For the soundness check for turtle creation, open inspect windows for several patches and check that the number of turtles is consistent with the value of the popn variables.

We can also use the inspect window to understand the effect of the `clear-all` primitive that was included as the first line in the setup procedure. Comment out that line, then press the setup button a few times with an inspect patch window open. You should see additional people being created. That is, the setup procedure is not resetting the model back to an appropriate initial condition. Uncomment the `clear-all` line and try again. The `clear-all` resets NetLogo to its default or empty state, as if the model had just been loaded. Among other things, it deletes any agents that have been created and sets all global and patch variables to 0 (their default value).

clear-all

The primitive `clear-all` is used to reset NetLogo to its default state, as if no code has been run. It is sometimes seen in code in its abbreviated form, `ca`.

Open an inspect window for an agent. In the code box at the bottom, type `set size 10` and then press Enter. The icon for the person you are inspecting will increase to size 10 in both the inspect window and the main World widget. This is an example of controlling NetLogo interactively, entering code for immediate implementation.

5.10 Colours

Unless otherwise specified, patches are coloured black and turtle agents are randomly coloured. The next model increment makes the interface informative by using colours effectively. First, however, we need to understand how NetLogo defines colours.

Open the tutorial model and press setup. Open an inspect window for a person and one for a patch. In the person window's code box, enter `set color red` (note the US spelling). In the patch's code box, enter `set pcolor 15`. What happened? What should happen is that both the person icon and the patch background change to red.

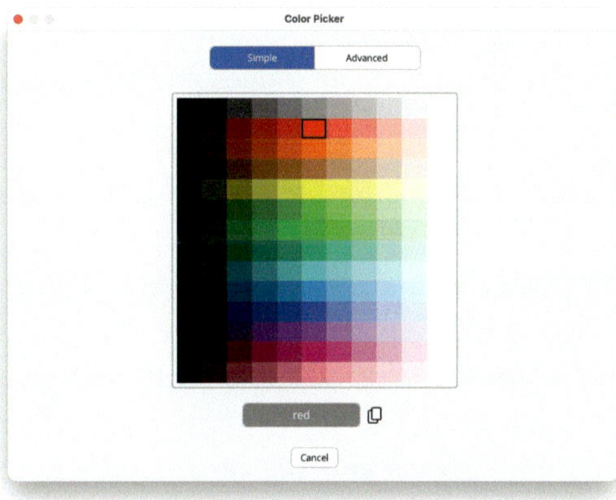

FIGURE 5.11: Basic colours in NetLogo.

Colours in NetLogo have equivalent numerical values, and the name and number can be used interchangeably. Valid colour numbers are from 0 to 140[4]. Each interval of length 10 refers to a broad colour such as red, which runs [10,20) or magenta, which runs [120,130). The lowest numbers in an interval are dark, and the higher numbers are pale. The mid numbers (that is, integers ending in 5) are the named colours.

The colour map is available from the NetLogo menu, at Tools > Color Picker. Select that to open the map displayed at Figure 5.11. Move the mouse pointer to different places and click; the name of that colour is displayed below the picker.

Switch to the Advanced view, set 'Input Format' to 'NetLogo number' and 'Netlogo Output Format' to 'NetLogo word'. Entering 15 in the input box will change the colour of the square to red and return the word 'red' in the output box. Drag the pointer around the window to pick other colours and display the number and word that can be used to select that colour. Note that the colour number does not have to be an integer. Now try the number 123. In the text box, this colour is described as 'magenta - 2', confirming the equivalence of a colour and number.

[4] Other numbers are not treated as errors; they are instead mapped back to the relevant range using modulo arithmetic.

The use of numbers for colours supports great flexibility in using colour to provide information about the model. We have already seen patches coloured to display whether food is available in the *Rabbits Grass Weeds* library model. The next step of the tutorial model is to use colour and other visual elements to communicate with the model user, highlighting important information.

5.11 Communication through User Interface Design

The tutorial model will use colour to indicate population density. The display takes advantage of the numerical equivalence of colours with the `scale-color` primitive, which rescales a numerical variable and assigns a colour based on that variable value.

scale-color

The `scale-color` primitive is used to map a value to a colour range. The full syntax is `pcolor` <colour-group> <number-to-map> <minimum-value> <maximum-value>. The minimum and maximum values can be swapped, and the colours will invert accordingly.

Follow the instructions for Snippet 1d-1 to create a procedure to colour the patches and call the new procedure as part of setting up the patches. We first create a local (temporary) variable named 'max-scale' and assign it a value higher than the maximum that will be required for scaling the colour. Remember that the `let` primitive is used to simultaneously create a variable and store a value in that variable (with `set` used to change it later if required). A logarithmic transformation is applied to the population values to reduce the extreme difference between the city patch populations and the other patch populations. The primitive `ln` is the mathematical function of natural logarithm. Multiplication by 1.3 ensures that even the highest values of population density do not generate colours that are too pale to be visible.

let

The primitive `let` is used to create a local variable and assign a value. The full command is `let` <variable name> <value>. The value can be specified explicitly (eg `10` or `"hello"`) or as the output of some piece of code. The variable created by `let` is temporary and will be deleted at the end of the block of code (delimited by [ ] square brackets) or procedure within which it is created.

The variable `pcolor` is the built-in colour variable for patches. So `set pcolor <...>` instructs the patch to change its colour to the stated value. In this case the stated value is derived from the `scale-color` primitive, with the logarithm of the value of the patch variable popn mapped to the `blue` colour range. In this example, the maximum possible popn value is 5 million (with `ln` of 15.4), which would result in a value of approximately 20 for max-scale. Overall, the population values are mapped to a scale between 3.2 and 20.05 and rescaled to the range 100 to 110, which is the `blue` range of the NetLogo colour system.

Snippet 1d-1: Colour patches

1. Add call to colour-patches procedure after the call to setup-patches (before the call to the make-people procedure) in the setup procedure.
2. In the (new) utility procedures section at the end of the code, create the following procedure to colour patches.
3. Check the code and Save the model.

Note: Shading is used for surrounding code that does not change, and the ellipses (...) indicates that there is code that is not shown.

```
to setup
  clear-all
  setup-patches
  colour-patches
  make-people
end

...

;-----------------------------------------------
; UTILITY PROCEDURES
;-----------------------------------------------

to colour-patches
  let max-scale 1.3 * ln max-popn
  ask patches
  [ set pcolor scale-color blue ln popn max-scale 3.2
  ]
end
```

pcolor

The built-in variable for patches to store their colour is named `pcolor`. It can be changed with `set pcolor` <value>. The colour value can be specified with either a name (eg `red + 1.3`) or the numerical equivalent (eg 16.3).

After completing this snippet, press the setup button and see what happens. The black background in the World is now blue, but the turtle icons are obscuring any variation in the shades, so the next step is to consider how to colour the turtle agents or people. The most relevant information about people is whether they have adopted protective behaviour and one option is to colour people by their protection status. However, the people on a patch are all in the same place, only the last created person is visible as it obscures all the others. It is therefore more sensible to hide the turtle agents entirely. There are two approaches to hiding turtle agents.

The first uses the hide-turtle primitive. Follow the instructions for Snippet 1d-2v1. After completing the snippet, press the setup button. The colouring to indicate population is now visible because the people agents are not visible. They are still created, though. You can see this by opening an inspect window for a person. The context menu is not available. Try it, and you will find that the menu does not include items for turtle agents. This is because hide-turtle completely hides the turtle from the World, rather than simply turning off the visible display. Instead, go to the Command Center (bottom of the interface, see section labelled (8) in Figure 4.1) and enter the code `inspect person 100` at the bottom space.

Snippet 1d-2v1: Initial approach: Hide agents

1. Change the make-people procedure by adding one line (within the code block run with sprout-people).
2. Check the code and Save the model.

Note: Shading is used for surrounding code that does not change and the ellipses (. . .) indicates that there is code that is not shown.

```
to make-people
ask patches
  [ sprout-people ...
    [ set protect? FALSE
      hide-turtle
    ]
  ]
end
```

hide-turtle

The command hide-turtle makes a turtle invisible in the world view. It is reversed with show-turtle.

A more useful approach (for our purposes) is to make the turtle icons so small they are invisible but not actually 'hide' the turtles. Edit the line you just added using Snippet 1d-2v2 to set the size of the turtles to 0. Make sure you delete the hide-turtle you added in Snippet 1d-2v1. Now press the setup button. The World looks the same, but you should also have access to the context menu to inspect people (click in multiple places until you find one).

Now that you can see the variation in population (blue shading for patches), does anything strike you as odd? Remember that the population was assigned randomly for each patch. This means that cities with high population can be adjacent to other patches with very low population. But this is not realistic. In the real world, big cities are surrounded by relatively high-density population areas and, similarly, low-population areas tend to cluster together. The next step is to better represent the patterns of the real world but without setting values that exactly match some real-world land mass. What is required is a smoother transition in population values instead of sharp differences at patch boundaries.

Snippet 1d-2v2: Revised approach: Minimise agents

1. Change the make-people procedure: replace **hide-turtle** with the new line.
2. Check the code and Save the model.

Note: Shading is used for surrounding code that does not change and the ellipses (...) indicates that there is code that is not shown.

```
to make-people
ask patches
  [ sprout-people ...
    [ set protect? FALSE
      set size 0
    ]
  ]
end
```

The `diffuse` primitive moves part of a variable's value to the surrounding patches. Complete Snippet 1d-3, which uses `diffuse` to smooth population values across patches. Each patch takes 40% of its own value of popn and shares it between the eight neighbouring patches. This means that the city centres create an initial ring of relatively high population values. The spreading operation occurs three time (due to the `repeat` primitive), so that there are three rings with gradually decreasing population values around each city. The city loses 40% of its population each time (leaving it with 0.6^3 or 21.6% of its initial value), but also gains some population from the sharing of its neighbours.

Snippet 1d-3: Diffuse population

1. Change the setup-patches procedure: add the new line after setting the population values.
2. Check the code and Save the model.

Note: Shading is used for surrounding code that does not change and the ellipses (...) indicates that there is code that is not shown.

```
to setup-patches
  ask patches ...
  ask n-of ...
  repeat 3 [ diffuse popn 0.4 ]
  set max-popn ...
end
```

diffuse

The `diffuse` primitive removes the specified proportion of a patch's variable and shares that portion equally between each of the eight neighbouring patches, adding the share to whatever value the receiving patch already has for that variable. The full syntax is `diffuse` <variable-name> <proportion-to-share>.

repeat

The `repeat` primitive instructs NetLogo to perform the following code block multiple times. The full syntax is `repeat` <number-of-occurrences> [code-to-repeat].

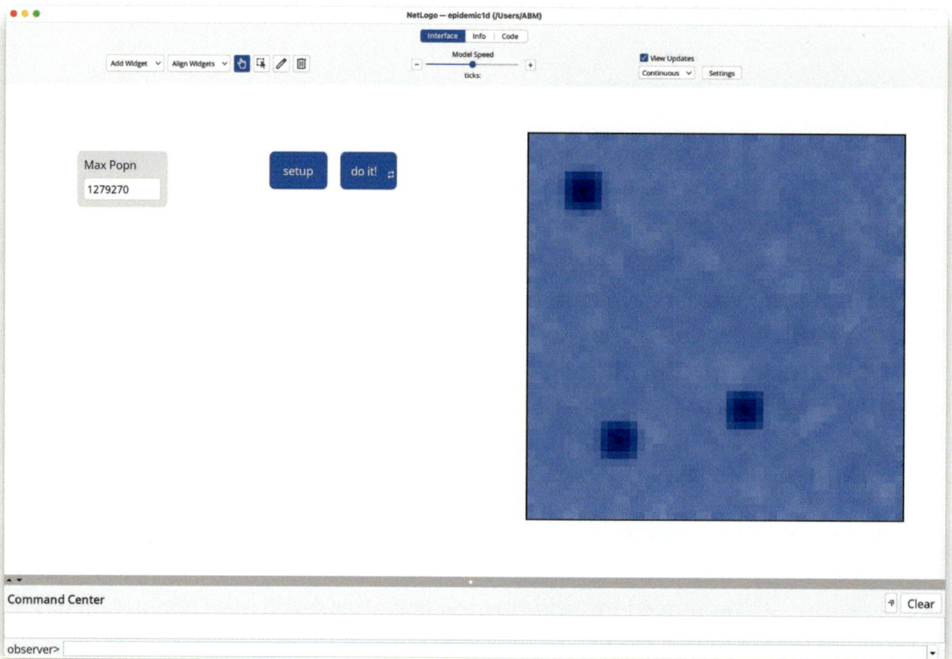

FIGURE 5.12: Screenshot of the interface after completing version 1d of the model.

Informative interface design is not only useful for communicating model results to the user, but it also reduces the need for explicit soundness checking as the code is constructed. The World displays the population values with the blue shading, making it clear whether the objective of smoother population has been achieved. But some checking is still appropriate. Inspect a dark blue patch and also patches in one direction outward; is the population value decrease reasonable? Also, is the new maximum population in the monitor at the expected level? Once you are satisfied that the population is spread throughout the world, model version 1d is complete. Save the model (Save item in the File menu).

5.12 Tutorial Progress Check

Version 1 of the model is now complete. The model creates turtle agents to represent people making decisions about their protective behaviour, and people are able to store their protective behaviour status. There is also a background population available to transmit the epidemic.

However, no decisions are actually being made. In fact, this model is not yet a simulation, as there is no passage of time and no processes are being modelled. Nevertheless, many important concepts have been presented.

5.12.1 Programming practices

At this stage, you should be much clearer about the iterative nature of developing code. You first think about what you want to achieve, then write code to achieve that objective, and then test the code to make sure it works. Each addition should be incremental, as small as possible while still being complete. The code must run so it can be tested.

NetLogo provides a syntax checker (*Check* button with tick icon). This will find problems of construction, where the entered code does not follow the syntax described in the NetLogo dictionary. For example, the `sprout` primitive must be followed by a number for how many turtle agents to create. If that number is missing from the code, the syntax checker will report an error. It will also identify errors such as missing `end` for a procedure, unpaired brackets, and variable or procedure names that are spelled inconsistently so cannot be found.

Each code snippet has included a reminder to apply the syntax checker and then save the model. Those reminders will no longer be included in snippet instructions; you should be checking and saving regularly as part of your programming habits.

This syntax checker does not help with logic errors. Common errors include creating a procedure but forgetting to include the call to run the procedure, and forgetting to set sensible initial values for variables. However, most logic errors are more subtle and can only be detected by explicit testing. So far, tests have involved checking that generated values are consistent with expected values, either displaying a variable value or by opening an inspect window.

Code should be modular; each procedure (bounded by `to` and `end`) is independent. Each procedure comprises primitives and calls to other procedures. A primitive is a defined word of the NetLogo language that instructs patches, turtles or links to do something or reports some information about these entities. Primitives are written by the NetLogo developers. In contrast, procedures are written by you, the model developer. Procedures that are invoked by higher-level procedures may in turn call other procedures even deeper into the levels of nesting.

In NetLogo, procedures can be included in the *Code* tab in any order. By convention, NetLogo has two main procedures: the *setup* procedure contains the code to initialise the model, and the *go* procedure contains the code to step through time and implement the processes. In this tutorial, the procedures are separated into groups in the following order: main control procedures (setup and go), implementation procedures, and utility procedures. Implementation procedures are called by *setup* or *go* to perform some task that is specific to the model. Utility procedures are calculations and other general purpose functions that could conceivably be used in other models.

Above the procedures, there are specific code sections that define the operating environment of the model by describing the types of agents and the variables that are to be available. The recommended order for these sections is:

- `globals` statement listing the global variables.
- `breed` statement identifying the types of turtle agents available to the model, paired with the appropriate ...
- `turtles-own` statement listing the variables available to the specified type of turtle. Note that these use the breed name rather than the word 'turtles' in the primitive.
- `patches-own` statement listing the variables available to each patch.

A key part of the developing the model is designing the interface. In NetLogo, the interface is automatically connected to the code. Interface widgets that are used to input variables (such as sliders) automatically define a global variable simply by their presence on the interface. Buttons are able to call procedures to run. And output widgets have a code box that contains the NetLogo code to identify the variable value or other information to report.

The World widget displays the model's world. It comprises a grid of patches, each of which represents a physical location and contains information about the environment. This world is populated by turtles, which represent decision-makers or actors in the processes to be simulated. Turtles come in different breeds.

Information in the model is stored in variables. Global variables contain information that is accessible to all model entities. There is only a single copy of each global variable, with each variable defined by either an interface widget or by being named in the `globals` list. Agent variables contain information about a specific patch or turtle agent. There is one copy for each patch or turtle, and they are defined by `patches-own` or `turtles-own` lists. Finally, local variables are temporary storage created as required for convenience and then destroyed at the end of the code block or procedure where they are created.

5.12.2 NetLogo language

Comments are used to describe the purpose of a procedure or a section of code to the human reader. They are created with the semicolon (;), with NetLogo ignoring the semicolon and the following text to the end of the line. Comments are also common to provide additional information about a variable when it is created, and multiple line comments are used as headings in the code to make navigation easier.

White space is important in NetLogo for both the computer and the human. For the code to be interpreted correctly, mathematical operators must have space on either side. This is additional to the usual requirement that variable names, primitives, and other code elements are separated with a space. However, space is not required around brackets. NetLogo does not care about the amount of space, and multiple spaces are treated the same as a single space.

Make your code neat

Neat code is easier to read and understand, which leads to fewer errors.

Space is also used to indent code, to make it readable for humans. NetLogo would interpret the code correctly if all the code was on a single line with a space between each element. However, it would be extremely difficult for a human to understand the code. Snippet 1c-2 demonstrates several good practices for code readability:

- Each instruction is on a new line.
- Each level of nesting (either the procedure delimiters `to` and `end`, or a block of code surrounded by square brackets) is indented, with further indenting at the next nesting level.
- The starting bracket for a code block is on a new line and lined up with the commands that implement the block, with the closing bracket on its own line and also lined up. This makes it easy to see that the brackets are paired correctly, and the limits of the code block.

Good indenting makes it easier to identify errors, update the code for new features, and adapt the code to different projects. As you have entered the code so far, you may have noticed that NetLogo does some indenting automatically. For example, if you end a line, the next line will have the same indent. Further, if you type a closing square bracket on a new line and then press 'Enter', the bracket will outdent, reversing the expected indent of the code block.

Finally, there are several explicit name requirements for both procedures and variables in NetLogo, with letters, numbers, and only certain special characters permitted. Names should be clear, informative, and concise. By convention, multiple word names are joined with hyphens (such as 'max-popn') and logical variables (and procedures) have a question mark at the end (such as 'protect?').

New NetLogo keywords explained in this model version:

- code structures: `to`, `end`, `clear-all`
- defining and creating turtle agents: `breed`, `sprout`
- creating and manipulating variables: `globals`, `patches-own`, `turtles-own` `let`, `set`
- built-in patch variables: `pxcor`, `pycor`, `pcolor`
- mathematics: operators (+ * − / ^), `random`, `ln`, `scale-color`
- miscellaneous: `diffuse`, `repeat`, `hide-turtle`

6

Model 2: Introducing Time and Space

Model version 2 implements the epidemic process, at which point the model is a simulation. Key to implementing a simulation is modelling the passage of time and changes in model states through time. This version also introduces some relatively simple interaction, the actions occurring in one patch affect other patches. Such influence arises because of the spatial relationships between patches, and this version will also introduce the key spatial primitive of `neighbors`.

6.1 Model Design: The Epidemic Process

Before trying to model the epidemic process, the process must be specified in more detail. The detailed design is then implemented in NetLogo code.

The tutorial model epidemic process is the simplest version of the SIR model with no births or deaths (Kermack and McKendrick, 1927; Diekmann and Heesterbeek, 2000), where S refers to susceptible, I to infected (and infectious), and R to Removed. Ignoring births and deaths essentially assumes that the epidemic process occurs in a much shorter time-frame than the time-frame for population change.

How does a person become infected? Consider a typical person's activities during a day (or some other arbitrary but fixed unit of time). They interact with some number of people sufficiently closely to permit transmission if one person is infected and the other susceptible. Of those contacts, only a proportion will lead to transmission because of other factors such as the viral load of the infected person and healthiness of the susceptible person. Let β denote the combination of the number of such contacts per unit time and the chance of infection given such a contact.

Once a person becomes infected, they have a fixed probability per unit time of recovering, with the probability denoted by γ (Greek letter gamma). Once recovered, they are immune from further infection. Note that the full SIR model allows a proportion of infected people to die rather than recover, so the state is referred to as removed from the population. These state changes and parameters are summarised at Figure 6.1.

With these definitions, the change in the population in each epidemic state can be described by a set of equations. The change in the susceptible population (ΔS, where the symbol Δ denotes 'change in') is the number of people changing states from S to I. The parameter β (Greek letter beta) denotes the number of people able to be infected by a single infectious person (in the given time period) if everybody else in the population is susceptible. But only a proportion of the contacts of each infected person are actually susceptible, that proportion is S/N. Further, the number of people instigating such infections is given by I. This leads to $\Delta S = -\beta I S/N$.

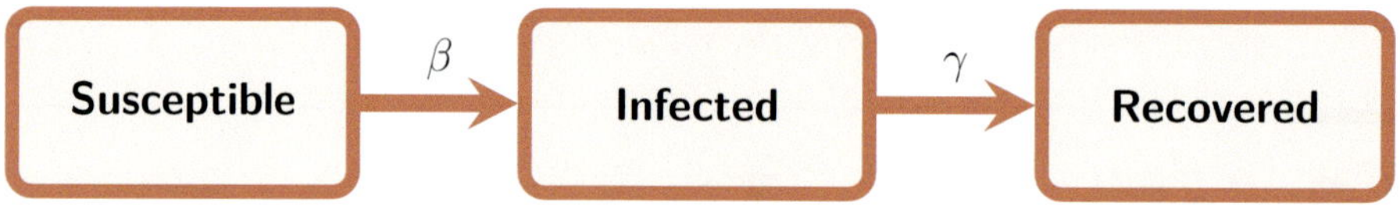

FIGURE 6.1: Epidemic state transitions for a population in the SIR model. There are no births or deaths, so the population is constant. For the given unit of time, the transition parameter β denotes the probability that a susceptible person becomes infected, and γ denotes the probability that an infected person recovers (and becomes immune).

Similarly, if γ is the probability of an infected individual recovering (in the given time period), then the infected population recovering is given by $\Delta R = \gamma I$. Finally, the change in the infected population is the gains from those becoming infected and the losses from those recovering. The relevant equations are therefore:

$$\Delta S = -\beta I \frac{S}{N}$$
$$\Delta I = \beta I \frac{S}{N} - \gamma I$$
$$\Delta R = \gamma I$$

Some mathematical manipulation (see Diekmann and Heesterbeek, 2000, or another epidemiology text) provides a more intuitive pair of parameters that can be used to derive β and γ. The duration of infection is $1/\gamma$ and the basic reproduction ratio is β/γ. The basic reproduction ratio is the number of new infections that would be generated by an average infected person if every contact they had was with a susceptible person.

Note that time is built in to these equations, change (Δ) in the population counts requires a time period over which the change occurs. Further, this process is complex, new infections arise from the interaction between the susceptible and infected populations. However, the representation is not agent-centric. The population is simply counted by epidemic state, and all people are treated as identical apart from their state. There is no concept of local interaction as occurred with the *Virus on a Network* library model, and individuals are not represented in the equations.

Refer to Figure 4.11, both protective behaviour and the efficacy of that behaviour also influence the epidemic process. However, those elements cannot be introduced until the model includes protective behaviour. This is an example of starting with the simplest model possible and then refining it gradually.

As part of the detailed design, the modeller should identify the requirements to implement the process to be represented and the expected behaviour of the model. This planning helps to organise the order in which to introduce new code, and the tests to conduct to assess whether the code is correct.

Looking at the equations, each patch needs a variable for the populations in each state and global variables are required for the state transition parameters β and γ. From general epidemic theory, an SIR epidemic has a characteristic pattern of infections, with incidence (new infections) and prevalence (current infections) increasing and then decreasing in the classic diffusion curve.

Eventually, the epidemic dies out because of the lack of accessible susceptibles. In addition, a basic check is that the total population in each patch does not change, just the allocation to different states.

6.2 Methods of Time Keeping

Just like any other aspect of the world that is to be included in the model, time must be represented in some way. Mathematically, time is continuous and many processes (including epidemic spread) occur in continuous time. This means that activity could occur at any arbitrary time.

One way to model continuous time is to track discrete events in a schedule (as with Discrete Event Simulation discussed at Section 1.4.2). These events can be generated by some function or triggered by other events. For example, in an epidemic, a person becoming infected is an event, which also triggers a later event of that person recovering. In this example, when the person becomes infected, the duration of the infection is calculated and the recovery event is added to the schedule to occur at the appropriate time. Other methods of working with continuos time include mathematical approaches such as differential equations.

In practice, however, agent-based models generally model time as a discrete variable. That is, the model includes an internal clock and each tick of the clock represents a specific period of time such as an hour or a week. The advantage of this method is that empirical observations about processes are typically recorded for some regular period of time, such as quarterly economic activity.

6.2.1 Watching the Clock - Ticks

In NetLogo, the `tick` command advances the internal clock, and the `ticks` primitive (note the 's' at the end) reports the number of ticks that have occurred since the simulation (and hence the clock) started. In addition, the `reset-ticks` command is used to prepare the clock, setting the tick counter to 0 (summarised at Figure 6.2).

This clock analogy relates to the discussion at Section 5.3 about initialisation (or start) in the setup procedure, and simulation (or continue) in the go procedure. The setup procedure contains all the instructions that must occur at time 0. Conventionally, the `reset-ticks` command is the final instruction. The go procedure contains all the instructions that must occur in a time interval of whatever duration is represented by a tick. Conventionally, the `tick` command is the final instruction.

The `ticks` primitive to report the current time-step is used when the time something occurs is important. For example, a seller might offer a discount to recent buyers and this rule would require a comparison between the `ticks` value of the most recent transaction for that buyer against the current `ticks` value. An example is included later in the model tutorial.

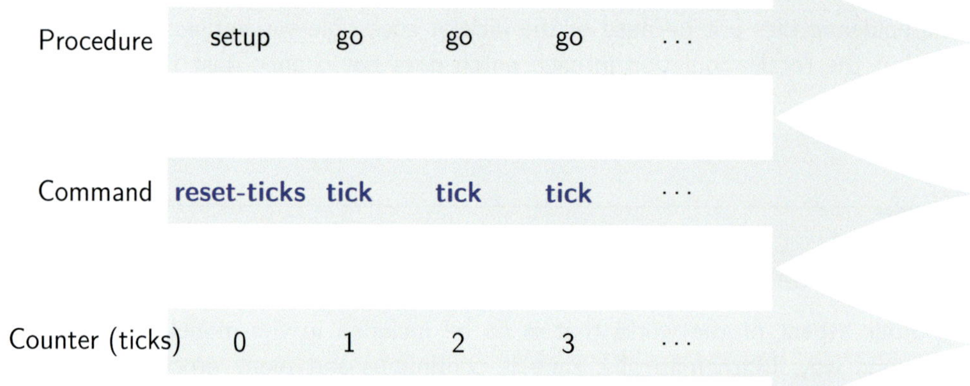

FIGURE 6.2: NetLogo represents time with a clock counter, accessed with the **ticks** reporter. The clock is initialised with the **reset-ticks** command, usually the last command in the *setup* procedure. It is incremented with the **tick** command, usually the last command in the *go* procedure.

6.3 Values for Agent Variables

The first step in spreading an epidemic is establishing some infected people to initialise the epidemic. Remember that the epidemic in this simulation is carried in the environment. This means that the patches must include variables for the number of people in each epidemic state.

6.3.1 Setting values

Follow the instructions for Snippet 2a-1 to create these variables and for Snippet 2a-2 to set appropriate default values (that is, before the epidemic is seeded). In these snippets, the letters 'S', 'I' and 'R' refer to the three possible epidemic states: susceptible, infected and removed (or recovered) respectively. The variable 'incidence' is the number of new infections. The population is initialised as entirely susceptible, so the values for I and R are set to 0 and the value for S is set to the patch population.

Initialise variables explicitly

It is good practice to explicitly set variable values even it they are to be set to 0 (the default value) so that the value is chosen consciously and to reduce the chance of leaving some variables out of the assignment process accidentally.

Remember that the snippet instructions will no longer include a reminder to check syntax and save the model. However, you should continue to check syntax and save after every snippet, and the model should pass the syntax check after each group of snippets with the same prefix (eg model 2a is not complete until after both snippets 2a-1 and 2a-2). The model may also pass the syntax check after adding one snippet from a group but this is not guaranteed.

The tutorial has already used the `set` primitive to assign values to variables (such as the popn variables in Snippet 1b-3) with limited discussion. This primitive is used to assign a value to

Snippet 2a-1: Define patch attributes

1. Modify patches-own list to add new variables for epidemic state counts.
Note: Shading is used for surrounding code that does not change.

```
patches-own
[ popn                 ; population of patch
  incidence            ; new infections this tick
  popn-S               ; susceptible population
  popn-I               ; infectious population
  popn-R               ; recovered population
]
```

an existing variable, regardless of the variable type (global, agent, or local) or the type of value (string, number, agentset, list). Here, `set` is being used to initialise the global variables. Note that some of these values are being set to 0 even though a variable is automatically assigned the value 0 when it is created. Systematically assigning values to all variables reduces the likelihood of missing a variable accidentally.

set

The primitive `set` is used to assign a value to a named variable. The full command is `set` <variable name> <new value>. The new value can be specified explicitly (eg `10` or `"hello"`) or as the output of some piece of code such as a calculation.

The new code in Snippet 2a-2 has the line `set popn round popn`. This will set the value of the popn variable to the integer that is closest to the current value of popn. This variable originally had an integer value for all patches, but the `diffuse` will have changed some of them to numbers with decimal places. Another primitive that forces numbers to be integers is `floor`, which was used in Snippet 1c-2.

round, precision, floor, ceiling

The primitive `round` finds the closest integer to the given number. If the number to be rounded is midway, then it increases. For example, `round 1.2` returns 1, `round 1.5` returns 2, and `round -1.5` returns -1. To round to a number of decimal places, use `precision` instead of `round`. The primitives `floor` and `ceiling` find the integer smaller than and larger than the given number respectively.

The changes introduced in these snippets are not visible in the model. After pressing the setup button, the model interface will be identical to that for model 1d (Figure 5.12).

Snippet 2a-2: Initialise patch attributes

1. Modify the setup-patches procedure to assign initial values for the patch variables.

Note: Shading is used for surrounding code that does not change and the ellipses (...) indicates that there is code that is not shown.

```
to setup-patches
  ...
  repeat ...
  ask patches
  [ set popn round popn
    set incidence 0
    set popn-S popn
    set popn-I 0
    set popn-R 0
  ]
  set max-popn ...
end
```

6.3.2 Seeding the epidemic

Now that there is a way to store information about the epidemic, the epidemic can be seeded. In the real world, epidemics require a critical mass of infected people to act as a source. To represent this in the model, a small number of high population patches must be selected as seeds, with a proportion of their population set to infected status. This is implemented in Snippet 2b.

Snippet 2b: Trigger epidemic

1. Create the start-epidemic procedure (in implementation procedures section of the code).
2. Add the call to the start-epidemic procedure into the setup procedure (before colour-patches).

```
to start-epidemic
  ask n-of 3 max-n-of 20 patches [ popn ]
  [ set popn-I round (0.005 * popn)
    set popn-S popn-S - popn-I
  ]
end
```

The order of the procedure calls in the *setup* procedure is very important. The *setup* procedure is mostly calls to other procedures, and those procedures are run from top to bottom as they appear in *setup*. Population values must be assigned for each patch before some of those patches are selected to seed the epidemic (since the selection depends on the population). This must occur before patches are coloured so that the colouring can display the location of the epidemic.

This snippet introduces two important primitives used for selecting a subset of model entities. `max-n-of` is used to identify the 20 patches with the highest values of the popn variable. From these 20 patches, 3 are randomly selected with `n-of`. The population state counts are adjusted in these three patches: the number of infected (popn-I) is set at 0.5% of the population (rounded to the nearest integer with `round`) and the susceptible population (popn-S) is reduced accordingly.

n-of

The full syntax is `n-of` **`<number> <agentset>`**. Selects a subset of specified size (number of entities) from the specified agentset, with each entity having an equal chance of selection. See also **one-of**.

max-n-of

The full syntax is `max-n-of` **`<number> <agentset> [<variable-name>]`**. Selects a subset of specified size (number of entities) from the specified agentset, selecting those with the largest values of the specified attribute variable. Ordering is random for those entities with the same value. This primitive has been presented as using a variable to order the entities, which is a common use. However, the primitive is more general, and any calculation that can be made for each entity can be used for ordering. The equivalent for selecting the entities with the smallest values is `min-n-of`. See also **max-one-of** and **min-one-of**.

6.3.3 Testing with the Command Center

What sort of soundness checks can be done to test this code? The code is intended to change the epidemic status counts in 3 of the 20 highest population patches, so one option is to open inspect windows at some of the dark blue patches until one has a non-zero value for popn-I. However, this is inefficient. Instead, we want to open a useful inspect window on the first attempt.

The *Command Center* (labelled (8) in Figure 4.1) has an area at the bottom to enter NetLogo code to be run immediately. You can use this to report the number of patches with a non-zero value for popn-I and open an inspect window for one of these patches.

Press the setup button to initialise the epidemic. Enter the code `count patches with [popn-I > 0]` into the input box. The output area repeats back the entered code (with `show` at the front) and reports the results of the `count` (see figure 6.3); as expected, there are three patches with popn-I values greater than 0.

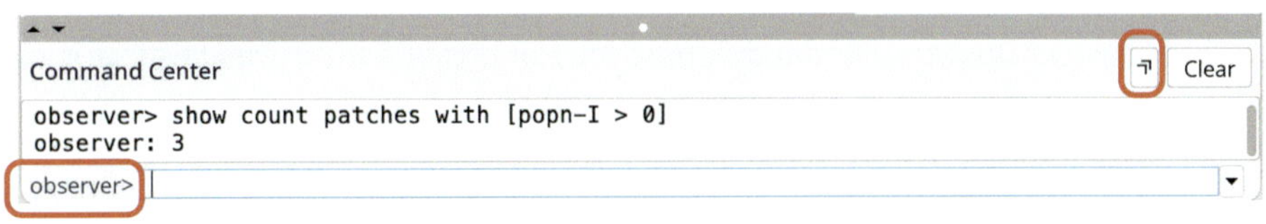

FIGURE 6.3: Command Center with results from code. The lines mark the context (perspective) drop-down box and the Command Center expansion toggle.

count

The `count` primitive reports the size (number of entities) in the specified agentset. The agentset can be specified explicitly by name (such as `patches`) or as the result from a piece of code.

There are other aspects of the *Command Center* to notice. Both the input area and the output have the word 'observer' at the start. This is to inform the user which perspective is being taken. Change the perspective to 'patches' using the drop-down box at the left of input area. Enter the code `set pcolor red` and you should see that all patches change their colour to red. Change back to the 'observer' perspective and enter `colour-patches`. This will run the procedure named colour-patches in exactly the same way as if it is called from the 'setup' procedure when the setup button is pressed. As only the entered procedure is run, the maximum population and the location of the cities is unchanged.

At the top right of the Command Center there is a double-ended arrow. Click on that and the text area will expand to take up the right half of the model interface. Click again and it will return to the small area at the bottom. The bar between the main interface and the Command Center can be dragged to change the size.

Now that you have explored the uses of the *Command Center*, we will use it to efficiently open a relevant inspect window. With the perspective set to 'observer', enter the code `inspect max-one-of patches [popn-I]`. What does this code do? First, it searches through the patches and finds the one with the largest value of popn-I (infected population) and then opens an inspect window for that patch. You should check that the values of the population in the different epidemic states look reasonable. In particular, check that popn-I is approximately 0.5% of popn, and that popn is the sum of popn-I and popn-S.

6.4 Conditional coding: Updating the colour scheme

So far the colour scheme displays population, but this can be extended to display the epidemic. Blue will continue to be used if the patch is mostly susceptible, with green if mostly removed (so the epidemic has passed). In addition, red will be used to indicate where the epidemic is active. This is implemented in Snippet 2c.

This snippet introduces all three versions of the conditional branching IF / THEN / ELSE. This is a fundamental construct of many programming languages and consists of a test, with one set of commands run if the test returns true and a different set run if the test returns false. It is essential for agent-centric behaviour because it translates the different characteristics and circumstances of each agent into potentially different actions.

The first use is the `ifelse` format. The test asks whether the susceptible population is larger than the removed population on that patch. If the test is true and there are more susceptible than removed, then the first code block is run (using a blue colour scale) and if it is not true then the second code block is run (green colour scale). The second use is the `if` format to test whether there is any infected population. The difference between `if` and `ifelse` is that nothing

Snippet 2c: Visualise epidemic

1. Amend colour-patches procedure to colour by epidemic state.

Note: Shading is used for surrounding code that does not change.

```
to colour-patches
  let max-scale 1.3 * ln max-popn
  ask patches
  [ ifelse popn-S > popn-R
      [ set pcolor scale-color blue ln popn max-scale 3.2 ]
      [ set pcolor scale-color green ln popn max-scale 1.5 ]
      if popn-I >= 1
      [ set pcolor ifelse-value (popn-I > 0.03 * popn) [16] [18] ]
  ]
end
```

```
set pcolor ifelse-value (popn-I
    > 0.03 * popn) [16] [18]
```

```
ifelse popn-I > popn * 0.03
[ set pcolor 16 ]
[ set pcolor 18 ]
```

FIGURE 6.4: **ifelse-value** can be used instead of an **ifelse** construction where the two paths are setting the same variable to different values. The left code box is the **ifelse-value** as included in Snippet 2c, and the right code box shows the **ifelse** with separate **set** commands. Note also the use of numbers for colours: 16 is a darker shade of red than 18.

happens if the test is false with `if`, the code block is simply skipped. Finally, `ifelse-value` is used if the two actions to be taken are simply to set the same variable to two different values. This has the same logic as `ifelse` but shorter code (see Figure 6.4).

ifelse, if, ifelse-value

The most general of these primitives is `ifelse`, for which the full syntax is `ifelse <condition> [<instructions-if-true>] [<instructions-if-false>]`. This allows the program to branch, running one set of instructions if the condition is `true` and a different set if the condition is `false`. If no code is to run when the condition is `false`, the primitive `if` can be used. If the two branches set the same variable to different values, then `ifelse-value` is the appropriate primitive.

After completing this snippet, press the setup button. Three patches in high population areas should be coloured light red (see Figure 6.5). Open an inspect window for one of these patches to confirm that the population values in each state are sensible. This example also demonstrates that soundness checking is much easier with an informative interface design.

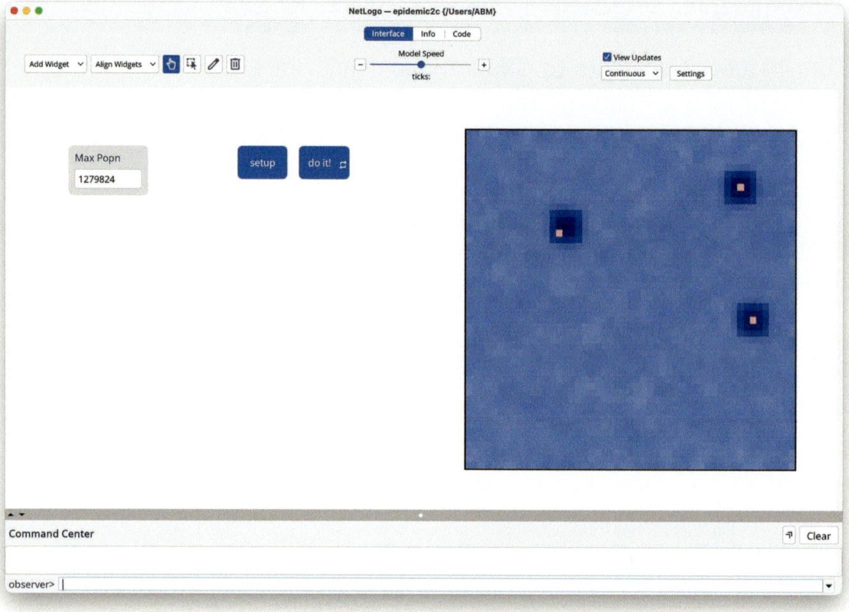

FIGURE 6.5: Screenshot of the interface after completing version 2c of the model.

6.5 The ask Command

One primitive has been used extensively already and not explicitly discussed. The `ask` command is essential in making NetLogo models. It implements the agent-centric approach by connecting the instructions to the model entities (turtles, patches, or links) that are to perform the instructions.

The model entities being asked are referred to as an agentset (discussed in more detail at Section 7.4). The agentset can be all the entities of a particular type, such as `ask patches [ ... ]` in the setup-patches procedure of Snippet 1b-3 or some identified subgroup, such as `ask n-of 3 patches [...]` in the same procedure. Agentsets can also consist of a single entity.

ask

The full syntax is `ask <agentset> [<instructions>]`. The `ask` command instructs each agent in the specified agentset to implement the code in the square brackets.

The `ask` command steps through each entity in the agentset and runs all the code for that entity before moving on to the next entity. Each `ask` moves through the entities in a random order. In a simulation, much of the code is run each time step (or tick), so any `ask` is also run multiple times. The random ordering avoids artefacts arising from the same agent always gaining any benefits of going first.

6.6 User Input Widgets

Examining the epidemic process equations in Section 6.1, there are two parameters that control the transitions between epidemics states: β and γ. Clearly, these could be added to the list of global parameters and assigned values with the `set` command. However, this approach would require the model user to edit the code to obtain different values. A better alternative is to assign them with interface controls, or widgets.

There are four types of widgets that allow values to be entered by the user into the model. A *slider* allows only numbers to be entered, and the specific number is selected by moving a slider. The developer sets the range of permitted values (minimum and maximum), which can be used to restrict entry to sensible values. A *switch* is used to enter a value that is either true (switch on) or false (switch off). A *chooser* (or drop-down box) is used to select between multiple pre-defined choices, which could be numbers, text, or some combination. Finally, an *input box* allows any value to be directly typed by the user, and the only restriction is the form of the value (such as number or string).

Regardless of the type of input widget used, the top of dialogue box has a space titled 'Global variable'. The text entered in that box must satisfy the variable naming conventions (see Section 5.5). This is because the widget creates a global variable with the given name. The variable is available to the code in exactly the same way as a variable listed in a `globals` statement (and will generate an error if it appears in both places). The value is assigned by the input rather than a `set` statement. Note that a `set` statement in the code will change the variable's value, both internally and on the interface. That is, the interface widget is 'live' throughout the simulation and can even be changed during a simulation run.

User Inputs

An important part of the interface design is identifying the key parameters that users will want to change. Those parameters should be included in the interface with sliders or other input widgets, and less important parameters included within the model code as global variables with sensible default values applied with a `set` command.

6.7 Making the Model a Simulation

The next few snippets concern the process of spreading the epidemic. This is the point at which the passage of time is introduced into the model. There is a conceptual leap involved, so it is important to think about how time is integrated with the code. For convenience, the key equations controlling the epidemic process are reproduced below. In a population of size N, the variables S, I, and R represent the number of people in each of the epidemic states, and Δ is the change in those populations in each time step.

$$\Delta S = -\beta I \frac{S}{N}$$
$$\Delta I = \beta I \frac{S}{N} - \gamma I$$
$$\Delta R = \gamma I$$

The first decision is the amount of time that a tick is to represent. For abstract models, the time step is arbitrary. However, if the simulation is representing a real-world system, then it will be calibrated to some sort of data that incorporates time.

The time step should be short enough that it is quicker than the fastest process being simulated. That is, agents should not be taking multiple actions within a single time step unless all the actions are part of one larger action (such as moving and eating in the *Rabbits Grass Weeds* model). But making the time step too quick is inefficient. A good rule of thumb is to set the time step so that a small but non-trivial proportion of the agents should change state (that is, take some action) each time step.

How long is a tick?

A `tick` should represent a time step during which a meaningful proportion of the agents change their behaviour or state.

For the tutorial model, a convenient time step is one day. This is because the time step should be shorter than the typical infectious period, which is about five days for an influenza-like epidemic.

6.7.1 Input epidemic parameters

The parameters controlling the epidemic are the variables β and γ. As they can both be derived from the average duration of infection and basic reproduction ratio, sliders will be constructed for these more intuitive measures and the epidemic parameters calculated within the model.

A slider is added to the interface in the same way as other widgets, like the button added at Section 5.4. Select slider in the Add Widget drop-down box and click on the interface where the slider should be located. Alternatively, open the context menu on the interface and select slider. The slider dialogue box will open.

You need to create two sliders and add them to the interface. The first is for basic reproduction ratio, which is denoted by R_0 in epidemiology and is the expected number of new infections arising from a single infected person in a susceptible population. Create a slider with variable name 'R0', minimum 0.25, increment 0.25, maximum 5, and value 2 (see Figure 6.6). Create a second slider with variable name 'recovery-period', minimum 1, increment 1, maximum 10, value 5, and units 'days'.

6.7.2 Stepping through time

The two parameters can be derived from the more intuitive measures added to the model with the sliders. If D represents the average recovery-period (in days, which is to be the time step

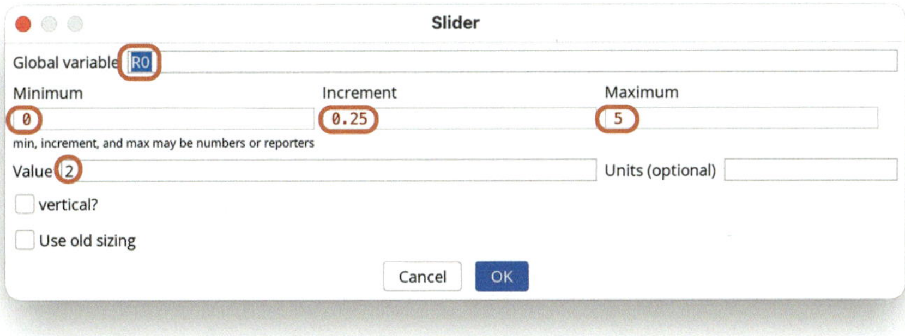

FIGURE 6.6: Screenshot of the slider dialogue box for the variable R0, which is the basic reproduction ratio. The top text box is used to enter the name of the global variable that is created by the slider. Other boxes are used to set the range and default value of the slider.

in this model), $\gamma = {}^1\!/_D$ (the probability of recovering on any day) and $R_0 = \beta D$ so $\beta = {}^{R_0}\!/_D$ (the expected number of people infected in one day by an infectious person). In addition, new infections (incidence) is given by: $\beta I\,{}^S\!/_N$.

Before implementing Snippet 2d-1, it is worth thinking about how you would translate these equations into NetLogo code. What order would you update these equations, and why is that order important? Are there any checks you need to make during the calculations?

The logic in this procedure is:

1. Calculate expected incidence first, before the S and I population counts are adjusted.
2. Make sure that incidence does not exceed the number of susceptible people available. This can occur because, conceptually, each infected person finds their own susceptible targets simultaneously so susceptibles can be infected multiple times. Reduce incidence if required.
3. Calculate the population to recover before the newly infected are added to the infected population so that the newly infected do not immediately recover.
4. Mathematically, the infected population would never reach 0 after it is non-zero because all the relevant equations simply multiply numbers together. Instead, both incidence and popn-I would approach 0. While rounding to nearest integer may resolve this and move them to 0, it is better to guarantee resolution by setting the infected population to 0 when it falls below 1.

The individual primitives in this procedure are relatively simple and have all been introduced already, comprising only `ask`, `if`, `let`, and `set`. Operating with information stored in variables, careful ordering of these simple commands is able to create sophisticated behaviour.

Note that rounding (to the nearest integer) is included in this code as it is not possible to have fractional people infected in the real world. However, this can introduce subtle errors when dealing with small numbers. This is the reason that popn-R is rounded rather than the local variable 'recovered'; the number of people recovering as the epidemic dies out will be less than 1, and rounding to 0 would mean that the last infected person never recovers.

Snippet 2d-1 implements the spread of the epidemic in a single time unit (day). The next step is to set up the passage of time, instructing NetLogo to repeatedly run the procedure to spread

Snippet 2d-1: Mechanism for epidemic spread

1. If you have not already done so, create two sliders as described in Section 6.7.1
2. Create the spread-epidemic procedure (a suitable location is as the first procedure in the implementation section).

```
to spread-epidemic
  ask patches
  [ ; calculate the number of new infections generated by the patch
    let beta (R0 / recovery-period)
    set incidence popn-I * beta * popn-S / popn
    ; update the population counts in each epidemic state
    let recovered popn-I / recovery-period
    if incidence > popn-S [ set incidence popn-S ]
    set popn-S popn-S - incidence
    set popn-I popn-I - recovered + incidence
    if popn-I < 1 [ set popn-I 0 ]
    set popn-R round (popn-R + recovered)
  ]
end
```

the epidemic. This is implemented in Snippet 2d-2. Once coded (and syntax checked), press the *setup* button and then the *do it!* button, some of the patches should change colour between pale red, darker red, and eventually green. After the simulation has run for awhile, press the *do it!* button again to stop it. The interface should look similar to the screenshot at Figure 6.7.

NetLogo links together the *do it!* button and the *go* procedure. The button calls the procedure when pressed. Checking the 'Forever' box means that the go procedure is called repeatedly until the button is pressed again. That is, a single press of the button requests multiple iterations through the go procedure (as depicted in Figure 6.2).

Code order

It is standard practice to locate `clear-all` as the first command in the setup procedure, `reset-ticks` as the last command in the setup procedure, and `tick` as the last command in the go procedure. As `reset-ticks` also initiates any plots, this order ensures that all the model entities are initialised before their values are plotted.

Each iteration through the *go* procedure first runs the spread-epidemic procedure, which updates the epidemic status counts for each patch. It then calls the colour-patches procedure, which we previously created so that the patches were appropriately coloured during initialisation (that is, the setup procedure) and amended in Snippet 2c to add colours for other epidemic states. Finally, it instructs NetLogo to increment the tick counter with the `tick` command. If `reset-ticks` command has not already occurred, then `tick` will generate an error. So the setup procedure is amended to include `reset-ticks`, initiating the clock as part of setting up the simulation run.

Snippet 2d-2: Step through time

1. Add **reset-ticks** at the end of the setup procedure.
2. Confirm that the 'Forever' checkbox is set for the *do it!* button.
3. Amend the go procedure.

Note: Shading is used for surrounding code that does not change and the ellipses (...) indicates that there is code that is not shown.

```
to setup
  ...
  make-people
  reset-ticks
end

to go
  spread-epidemic
  colour-patches
  tick
end
```

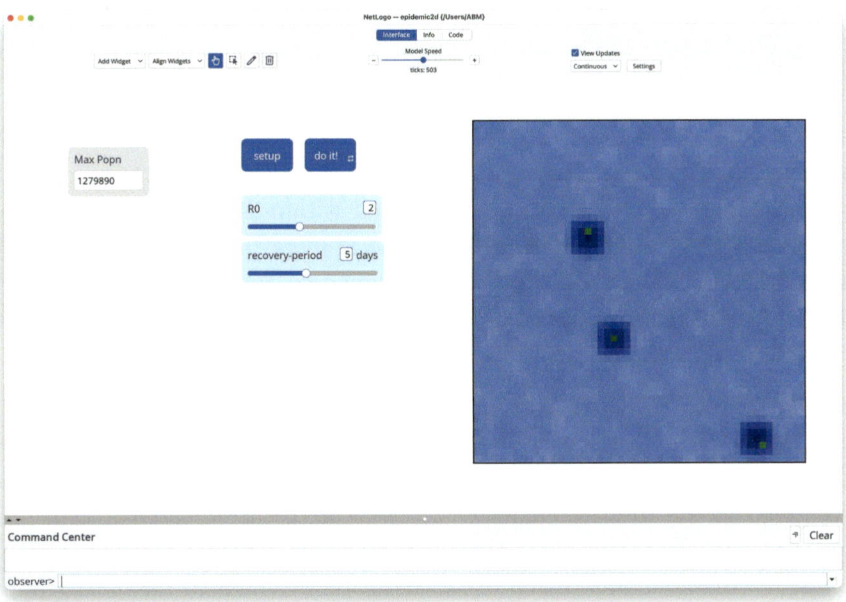

FIGURE 6.7: Screenshot of the interface after completing version 2d of the model. Notice the tick counter (below the speed slider) reading 503 in this example and the green colour of the patches where the epidemic occurred.

tick, ticks, reset-tick

The `tick` command increments the tick counter. The number of ticks is used to represent the passage of time and can be accessed with the `ticks` reporter. See also `reset-ticks`, which is used to initialise the tick counter.

To understand what is happening and as a soundness check, you can edit the *do it!* button to uncheck 'Forever', then press the setup button once and then the go button. Open an inspect window for one of the red patches. Each press of the *do it!* button will run through the go procedure once, updating the epidemic state population counts and incrementing the tick counter. Watch the incidence and population values in the inspect window as you press the button. See if you can use the equations to calculate the expected values for the epidemic state populations for the next 'go' step, then press the 'do it!' button and see if you are correct. Try it again with different parameter settings; for example, you would expect a more intensive epidemic with a higher value of R_0 and a shorter recovery-period. When you have finished, check 'Forever' again for the *do it!* button.

What did you notice while running the simulation? First, you had to stop the simulation by pressing the go button, even though the epidemic had run its course. Second, the epidemic didn't spread geographically, it was trapped in the patches where it started. Fixing these two issues, and plotting some simulation results, are the remainder of the tutorial model version 2. However, some additional programming concepts are required first.

6.8 Commands, Reporters and Procedures

There are several types of keywords in the NetLogo language. At the first level, NetLogo distinguishes between keywords and primitives. Keywords establish the NetLogo program rather than run it, and include `globals`, `patches-own`, `to`, and `end`. In contrast, primitives are the NetLogo words that are used within procedures.

Those primitives that are used to obtain information are referred to as reporters. Important reporters include `n-of` and `patches`, as well as several that will be introduced in the next version of the model. Other primitives that can be used within procedures are referred to as commands. Commands are primarily used to instruct one or more turtles or patches (or links) to do something. Important commands include `ask`, `set` and `ifelse`. Looking at this paragraph, you will see that all the reporters are purple and commands are blue, this is part of the automatic formatting in the NetLogo code area. And keywords are formatted in green.

In the same way that the primitives are split between reporters and commands, two types of procedures are available. A reporter procedure runs the contents and then provides some piece of information back to the procedure that called it. Command procedures implement actions. All the procedures included in the tutorial model so far are command procedures, and have been delimited with `to` and `end`. Reporter procedures are delimited with `to-report` and `end` and must include a `report` command identifying the value to be returned to the calling procedure.

to-report <procedure-name> <instructions> report end

A reporter procedure is a block of instructions that concludes with calculating some piece of information and returning that value to the procedure or model entity that called the reporter procedure. The keywords `to-report` and `end` mark the beginning and end of the block. In addition, a `report` command must be included within the block to identify the information to provide to the caller.

Command or reporter?

It is not always obvious whether to construct a procedure as a command procedure (using `to`) or as a reporter procedure (using `to-report`). A good rule of thumb is that, if the procedure changes anything in the model, it should be a command procedure. Reporter procedures should be completely passive, accessing information but not affecting the model in any way.

6.8.1 Ending the epidemic

There is no point in continuing the simulation after all the infected people have recovered. Snippet 2e-1 and Snippet 2e-2 combine to test whether the epidemic has ended and, if so, to override the 'Forever' checkbox to end the iterations through the go procedure.

The first step is to check whether the epidemic is over. Look at the code at Figure 6.8 and step through the logic (do not create the code, just read it):

1. A reporter procedure named 'all-done?' is constructed with a `to-report` declaration. Note that, like the people attribute 'protected?', the procedure name ends with a question mark. This is because the information that will be reported is the value `TRUE` or `FALSE`.
2. A local variable called 'test-result' is created to store the value that is to be reported at the end of the procedure.
3. A local variable called 'total-infected' is created. The total of all the patches' infected population values (popn-I) is calculated and assigned to this variable.
4. If the total is more than 0 (tested with `ifelse`), the the epidemic is still active and the value `FALSE` is assigned to the variable 'test-result'. Otherwise, the value `TRUE` is assigned.
5. The value of test-result (`TRUE` if the epidemic is over, and `FALSE` if it is ongoing) is provided to the procedure or model entity that called the 'all-done?' reporter procedure.

Add Snippet 2e-1 to the model. This implements the same logic as the code in Figure 6.8, but in a single line. The primitive `ifelse-value` can be used instead of `ifelse` because the action for either result is to `set` the same variable with different values. More subtly, the two values calculated in this procedure (named 'test-result' and 'total-infected' in Figure 6.8) are not explicitly stored in named variables, but are simply used as soon as they are calculated.

In Snippet 1b-3, the maximum of the population values over patches was calculated and assigned to the global variable 'max-popn'. This procedure takes a similar approach, calculating the `sum` (total) instead of the `max` (maximum). Both `sum` and `max` are mathematical primitives that

```
to-report all-done?
  let test-result 0
  let total-infected sum [popn-I] of patches
  ifelse total-infected > 0
   [ set test-result FALSE ]
   [ set test-result TRUE ]
  report test-result
end
```

FIGURE 6.8: An expanded equivalent to the code in Snippet 2e-1 to assist with understanding.

Snippet 2e-1: Test if epidemic completed

1. Create the all-done? procedure in the utility section.

```
to-report all-done?
  report ifelse-value (sum [popn-I] of patches > 0) [FALSE] [TRUE]
end
```

operate over a list. A list is a particular type of data structure, where multiple data items (numeric values, text strings, turtles, other sub-lists) are stored in a single variable in a fixed order. While lists are useful, they are outside the scope of this tutorial and will be discussed at Section 12.2. What is important here is that the primitive `of` returns a list. In this snippet, the list contains the values of the popn-I variable for each patch (in random order). That list is used as soon as it is created by the `sum` operator, without being explicitly stored; the variable is implicit.

of

Model entities automatically know the values of their own variables and global variables, and turtles also have access to the variables of the patch where they are located. The reporter `of` is used to access variables owned by other entities; it identifies the owner of the variable to be accessed. The full syntax is **`[<reporter>] of <agentset>`**.

6.8.2 Testing truth values

Boolean variables are those that only take the values `TRUE` or `FALSE`. Unlike some other languages, NetLogo does not actually know which variables are boolean. The convention of using a question mark at the end of the variable name is for the humans reading the code. The values of `TRUE` and `FALSE`, however, are special and are recognised by NetLogo. Note that `TRUE` is not the string "TRUE", it is a specific truth value. Furthermore, capitalisation does not matter, any mix of upper and lower case that spells the word 'true' or 'false' is interpreted as `TRUE` or `FALSE` respectively.

Truth values

TRUE and FALSE are special values recognised by NetLogo. They are created by logical tests or conditions and can be manipulated with logical operators.

If you want to test whether a variable has a specific value, then the test would have similar syntax as `if varname = "xyz" [ ]`. Further, to test whether it has any value other than the specified one, the code would be `if varname != "xyz" [ ]` (the combination '!=' is the operator for 'not equal to'). Equivalently, for a variable being used to store boolean values, the syntax would be `if varname? = TRUE [ ]`.

However, if the test is of a truth value (that is, either TRUE or FALSE), it is sufficient to drop the '=' part of the test. The code `if varname?` will report TRUE or FALSE as appropriate, and return an error if the variable 'varname?' has any other value (such as 0 or the string "TRUE").

But what if you want to test for FALSE? Just as there are mathematical operators to manipulate numbers (such as +), there are logical operators to manipulate truth values. In particular, `not` converts TRUE to FALSE and FALSE to TRUE. So, `if not varname?` would resolve to TRUE only when the value of 'varname?' is FALSE. Other common logical operators are `and` (resolves to TRUE when both values or tests are TRUE) and `or` (resolves to TRUE when either value or test is TRUE).

Instead of explicit variables, Snippet 2e-2 tests the truth value of the procedure created in Snippet 2e-1 to break out of the 'go' loop once the epidemic has run its course. Each loop through the go procedure, the all-done? procedure is called and, if that procedure returns TRUE, then the condition is met and the `stop` command breaks NetLogo out of the forever instruction issued by the button for the go procedure.

Snippet 2e-2: End simulation

1. Amend the go procedure to call the all-done? procedure as its first line.

Note: Shading is used for surrounding code that does not change.

```
to go
  if all-done? [ stop ]
  spread-epidemic
  colour-patches
  tick
end
```

stop

`stop` forces the model entity (including the observer) to terminate the code block that it is running. If implemented within a code block delimited with square brackets, the effect is the same as jumping to the closing bracket (and then continuing). It can also be used to terminate a procedure called by a button with forever checked.

This snippet completes model 2e. No screenshot is presented as it is identical to Figure 6.7. However, the simulation should stop running automatically. Press the *setup* button and then the *do it!* button to check the termination.

6.9 Spatial Awareness

The spatial infrastructure of a model is created automatically by the NetLogo system. This comprises the grid of patches that make up the World, and a co-ordinate system that allows each turtle to recognise its location. The NetLogo language has many primitives that take advantage of this spatial awareness to interpret concepts such as distances, directions and neighbours.

6.9.1 Co-ordinate system

Patches are identified by a pair of numbers such as `patch -5 3`. The first number is the `pxcor` (or patch x co-ordinate where x refers to horizontal or left / right) and the second number is the `pycor` (up / down). The numbers refer to the number of patches away from `patch 0 0`, which is in the centre of the world by default[1]. So this example would be the patch 5 patches to the left and 3 above the centre patch. The variables `pxcor` and `pycor` for each patch are automatically created with the World.

Similarly, turtle agent locations are specified with a pair of variables named `xcor` and `ycor`. Turtle positions are continuous, for example `xcor` of -5.2345 and `ycor` of 2.6789. The co-ordinate system for turtles is the same as the co-ordinate system for patches (despite the different names for the agent variables).

The integration of the co-ordinate systems means that a patch is exactly 1×1 spatial units in size. The patch that a turtle occupies is the one identified by rounding the co-ordinate values for the turtle, with 0.5 always rounded up. For example, the turtle with `xcor` of -5.2345 and `ycor` of 2.6789 would be on patch -5 3. Similarly, `patch 0 0` covers $-0.5 \leq$ `xcor` < 0.5 and $-0.5 \leq$ `ycor` < 0.5.

This built in co-ordinate system allows spatial concepts such as distance and directions to be meaningful. Turtle agents also have a heading automatically, even if they are not moving; it is a number from 0 to 360 (degrees) with 0 as 'up'. The spatial structure also supports concepts such as 'here' (referring to the same patch) and, for turtles, 'forward' and 'the patch in front of me'. For the tutorial model, the turtle agents will be stationary. Movement is discussed in Chapter 11.

[1] The position of 0, 0 can be moved with the Settings dialogue box

```
to colour-neighbours
  clear-all
  ask n-of 3 patches
  [ ask neighbors
    [ set pcolor yellow
    ]
  ]
  ask n-of 3 patches
  [ ask neighbors4
    [ set pcolor cyan
    ]
  ]
end
```

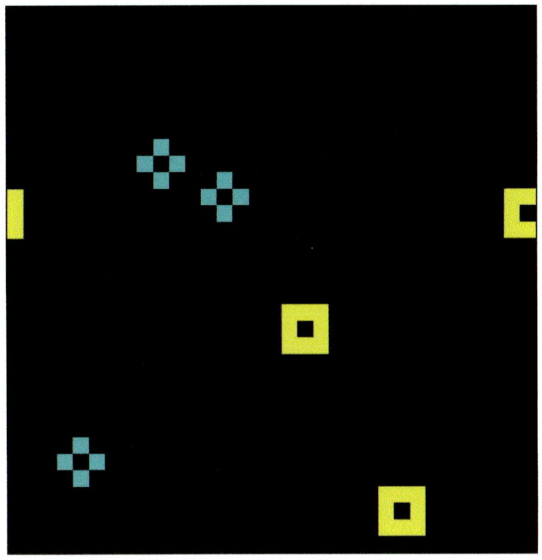

FIGURE 6.9: Demonstration of the **neighbors** (8 patches) and **neighbors4** (4 patches) reporters. Each of three randomly selected patches instruct their **neighbors** to turn yellow, and another three instruct their **neighbors4** to turn cyan. If the edges of the World wrap, the neighbourhoods will also wrap.

6.9.2 Neighbourhoods in NetLogo

NetLogo has two primitives to identify neighbouring patches, depending on whether you want all eight with a common corner (`neighbors`) or only the four with a common edge (`neighbors4`). Turtles are also able to use these primitives to refer to the patches bordering the patch where the turtle is located. These are demonstrated at Figure 6.9, with the code on the left generating the model world on the right. You can test this yourself by starting a new model, entering the procedure into the code tab, and then running the procedure by typing the procedure name into the Command Center.

neighbors

The eight patches adjacent to a patch are referred to as the patch's `neighbors`. A turtle can also access the `neighbors` of the patch where it is located. See also `neighbors4`.

6.9.3 Spatial epidemic

In order for the epidemic to spread spatially, some proportion of new infections must be created on neighbouring patches. The first step is to create a variable to store that proportion. That could be created on the interface with a slider or other input widget. However, to reduce clutter, the interface should be reserved for important settings that a user may wish to adjust to test out different scenarios.

Instead, the variable will be created as a global variable within the code. Creating it this way means that it will have a value of 0, so the value must also be explicitly set (or initialised) to something more appropriate. Both these tasks are implemented in Snippet 2f-1.

Snippet 2f-1: Establish travel

1. Create new global variable 'travel-rate' (by adding it to the **globals** list at the top of the code.
2. Create the initialise-globals procedure. While specific location is not important, it must be after the breed statements and other declarations (a suitable location is immediately above the setup procedure).
3. Amend the setup procedure to call the initialise-globals procedure as the first line after **clear-all**.

```
to initialise-globals
  set travel-rate 0.1
end
```

The `globals` statement declares the existence of this new variable. The initialise-globals procedure uses `set` to assign a value. Setting all the global variable initial values in a single procedure allows them to be easily found and changed if necessary, even by a model user. The call to that procedure must occur before any of the variables are used by other procedures, otherwise the value 0 will be used instead. Incorrect global values can introduce subtle logic errors that are difficult to detect and NetLogo will not report an error (unless, for example, the variable has a value of 0 and is the denominator in a division).

Setting global variable values

Setting the values for all the global variables in a single procedure reduces the chance of accidentally neglecting to set the value for one of them. It is good practice to include the variables that will have values of 0, even though NetLogo would set the value to 0 automatically. The procedure to set the values should be called as the first procedure to run after the **clear-all** command.

Snippet 2f-2 uses the new global variable to allocate some of the change in total infected population to neighbouring patches rather than just the patches where the infections are generated. The patch does this by randomly selecting one of the eight patches surrounding it and changing the incidence value in that patch. However, a patch could receive new infections from more than one patch, so care must be taken with the sequencing of actions.

The various sources of new infections must be summed within each patch to calculate the total incidence for that patch, new cases from the infected population within the patch but also those from neighbouring patches. This means that all patches must have the opportunity to generate new infections before the value of new infections is used to update patch epidemic status counts. In addition, since incidence is being cumulated rather than calculated, it must be explicitly set to 0 at the beginning of each time step so that only the new infections in the new time step are added together. As a result, the spread-epidemic procedure must have multiple `ask patches [ ]` code blocks so the first can complete calculations for every patch before using the results of those calculations in the second block.

As well as the already discussed `neighbors` primitive, this snippet introduces `one-of`. This is similar to `n-of`, discussed at Section 6.3.2, but randomly selects only one member of the specified set of agents (patches, in this case).

This snippet also introduces `myself` but this is a difficult primitive to understand, and some further work is required before it can be fully explained. At this point, it is sufficient to recognise that it is referring to the patch that did the asking.

Snippet 2f-2: Effect of travel

1. Amend epidemic-spread procedure to distribute the new infections.

Note: Shading is used for surrounding code that does not change and the ellipses (...) indicates that there is code that is not shown.

```
to spread-epidemic
  ask patches [ set incidence 0 ]
  ask patches
  [ ; calculate number of new infections generated by the patch
    let beta (R0 / recovery-period)
    let new-cases popn-I * beta * popn-S / popn
    ask one-of neighbors [ set incidence incidence +
                           travel-rate * [new-cases] of myself ]
    set incidence incidence + (1 - travel-rate) * new-cases
  ]
  ask patches
  [ ; update the population counts in each epidemic state
    let recovered popn-I / recovery-period
    ...
end
```

After implementing Snippet 2f-2, press the *setup* and then *do it!* buttons to see the effect of the changes. You should see the epidemic spread throughout the world, as shown in Figure 6.10.

6.10 Output Widgets for Model Results

So far, model reporting has focused on diagnostic uses of widgets (such as the maximum population monitor that provided a soundness check as the code change) and ensuring that the world displays the most useful information in informative ways (colours). However, there are several widgets available to report information to help users to understand what is happening in the simulation. Output widgets are constructed in the same way as the input widgets (see Section 6.6), by selecting a type from the widget drop-down box, placing it on the interface, and then entering commands and other information into the dialogue box.

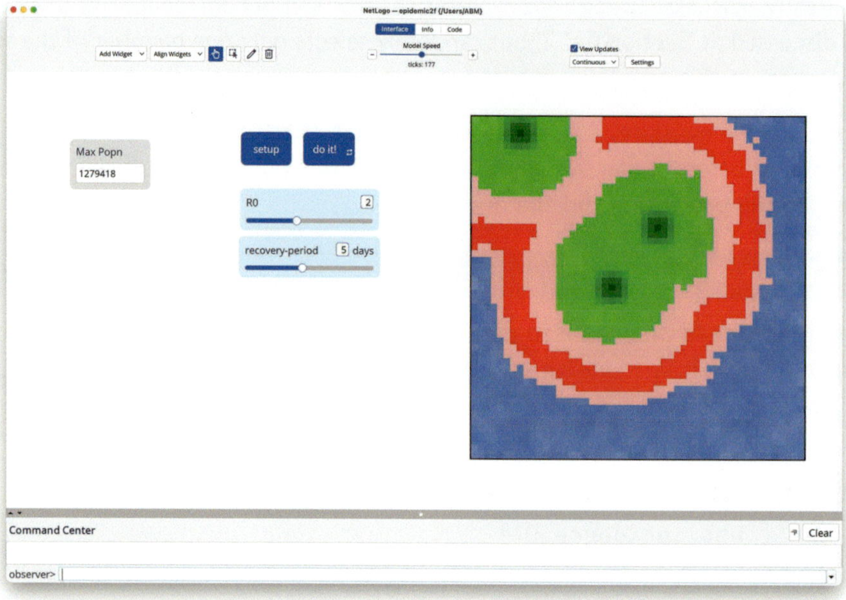

FIGURE 6.10: Screenshot of the interface after completing version 2f of the model. The epidemic should spread throughout the world, with a darker red epidemic frontier and patches turning green as the epidemic has passed.

Typically, a NetLogo model will include a line chart style of *plot* that displays some model information on the y-axis against time (ticks) on the x-axis. The world view presents the state of the simulation at a specific tick (describing what is happening 'now'). However, simulations represent a process, and it is often important to summarise how some aspect of that process changes over time.

Other plots are available for other types of information. Histograms are demonstrated at Snippet 3b-3, and it is also possible to plot two values against each other as they change over time. A *monitor* is used to display a single value, such as the contents of some variable or the results of a calculation. A *note* places fixed text on the interface, such as a heading. Finally, text can be generated by the model and written to an *output* area on the interface.

6.10.1 Reporting the epidemic

Key information about an epidemic is the incidence (new infections) and prevalence (current infections). A simple epidemic follows the classic diffusion curve, with an initial increase, slowing, and then decrease. Plotting these curves will provide further verification that the epidemic is behaving as expected. It will also show the way in which the epidemic behaviour responds to changes in epidemic parameters more easily than can be seen in the world view because the history for the simulation is visible.

When comparing multiple epidemics (for example, the effect of changes in input parameters) it is easier to remember numbers than curves. A monitor will be added to the interface to report the impact of the epidemic, the total population ever infected. You should think about how to calculate these values before moving on.

It is likely that you will need to move the existing widgets to make room for the plot. Select a widget (popup menu at the widget) and the border will be visible, you can then drag it with the mouse. Selecting a widget also allows it to be resized by dragging an edge or corner. If you want to move multiple widgets, use the 'select interface items' tool (in area (2) of Figure 4.1).

Snippet 2g describes the plot and monitor to be added to the interface. The plot has two pens (different coloured lines), one for incidence and the other for prevalence. That code is entered into the plot dialogue box, not the code tab of the model. Figure 6.11 displays the relevant plot dialogue box with the code and other details completed.

Similarly, the monitor code in Snippet 2g is for the Impact monitor, not the code tab. The monitor will display the total of the popn-I and popn-R values (all the people currently infected or ever infected, respectively). Set decimal places to 2.

Snippet 2g: Summarise epidemic progress

1. Delete the maximum population monitor (popup menu on the widget).
2. Create a plot titled 'Epidemic' with pens for 'Incidence' and 'Prevalence'; the code for each pen is provided (don't include the comment).
3. Create a monitor titled 'Impact' with the provided code for the report box.

```
; code for incidence and prevalence pens respectively
plot sum [incidence] of patches / sum [popn] of patches
plot sum [popn-I] of patches / sum [popn] of patches

; code for impact monitor report box
(sum [popn-I] of patches + sum [popn-R] of patches) / sum [popn] of
   patches
```

After implementing Snippet 2g, press the *setup* and then *do it!* buttons to see the effect of the changes. As the epidemic spreads, you should see the plot curves increase and then decrease (see Figure 6.12), potentially with additional peaks as the epidemic reaches areas with high population density.

Try changing the epidemic parameters and seeing the effect on the epidemic. You should find that a larger value of R_0 and / or shorter recovery period increases the intensity of the epidemic (steeper curves) and also the impact (total population affected). However, regardless of the parameter values, the impact will not reach 1, some population will remain susceptible. What happens if you reduce R_0 to below 1? This means that each infected person infects less than one other person on average, so the epidemic dies out very quickly. Because of the spatial effects (where infected people are concentrated), this dying out will actually occur at some value greater than 1.

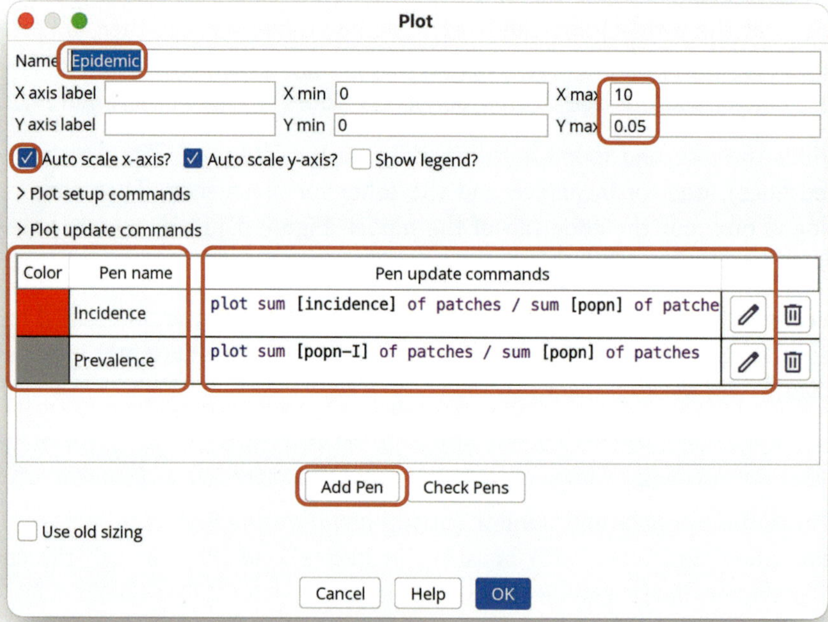

FIGURE 6.11: Screenshot of the plot dialogue box for epidemic progress. The top text box is used to enter the title of the plot. To the right are the (starting) limits of the axes. The X-axis is for ticks, so the default maximum value of 10 is reasonable. However, the Y-axis is proportion of the total population, and a small number is required to make sure the plot curves are visible. The 'Auto scale?' checkbox allows these axis limits to increase as required, and is essential for the x-axis when plotting against ticks. The button at the bottom of the dialogue box creates extra pens. Each pen is a line with a colour (changed by clicking on the coloured area), name, and code (commands). If there is insufficient space in the code box, press the pencil icon button at the right to access a further dialogue box.

6.11 Tutorial Progress Check

Version 2 of the model is now complete. The model represents the process of an epidemic spread and is therefore a simulation. There are also aspects of an agent-based model; changes on each patch are influenced by the surrounding patches and patches have different characteristics. However, there is no sense of agency, taking actions, or decision making; the patches are simply responding to their environment. It would satisfy only a broad definition of agent-based modelling.

Space and time are critical elements in agent-based modelling introduced in this model version. NetLogo is equipped with a spatial structure (and co-ordinate system) and a counter for the passage of time, as well as a range of keywords to access and control time and space.

Time is necessary for any representation of a process. In NetLogo, time is measured in arbitrary units referred to as ticks. In the tutorial model, each time step represents one day and the `tick` command advances the time counter one simulated day, but some other interval could have

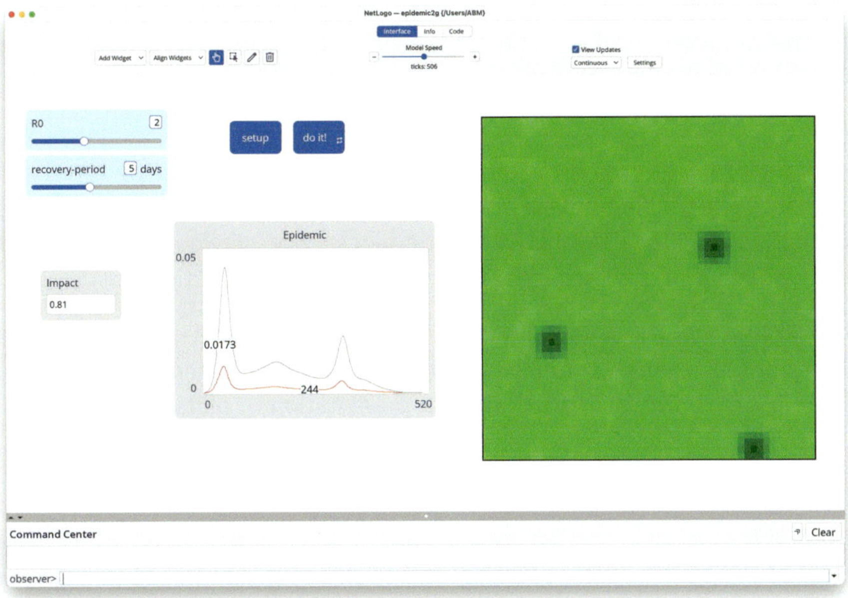

FIGURE 6.12: Screenshot of the interface after completing version 2g of the model, with the plots and impact monitor. This screenshot was captured at the end of the simulation, when the epidemic is over, to show the classic shape of the epidemic curves.

been used. Such connection to a real world time unit is necessary where empirical data must be translated for the model, so that appropriate parameter values can be assigned.

Space is necessary to represent the environment, an important source of influence and resources in many agent-based models. The spatial primitive `neighbors` uses the spatial structure inherent in the NetLogo platform to implement local actions.

This version also introduced conditional behaviour and boolean variables. These both require that NetLogo recognise `true` and `false`, not simply as words but as truth states. Conditions are used to control the flow of the code, so that different commands can be implemented in different situations. So far, this has only been used for visualisation, to colour the world. In the next version of the model, conditional code will allow agents to make decisions that reflect their own characteristics and circumstances, the essence of agent-based modelling.

Also formally presented was the fundamental command `ask` to pass instructions to specific agentsets, which had been used in version 1 code snippets without comment. Detailed discussion is deferred until agentsets have also been presented.

6.11.1 Programming practices

In this version of the model, the programming practices introduced in the previous version were reinforced. Each new snippet added as small a change as possible and both the syntax and logic were tested before making further changes.

The code is modular, with each separate procedure implementing a specific action and potentially calling other procedures for subsidiary tasks. The *go* procedure comprises a series of calls to the procedures that implement the epidemic process. This is equivalent to the *setup* procedure controlling the model initialisation. The main control procedures are placed near the top of the code tab to make it easy to see this highest level ordering of tasks.

Control flow with `if` conditions is a standard programming approach. Other control flow structures included in NetLogo are `while` and `loop`, neither of which is used in the tutorial. They all follow the same structure, the code block is run if some condition is satisfied. Unlike `if`, `while` and `loop` allow multiple iterations through the code block. The previously presented primitive `repeat` controls multiple iterations through a code block without conditions.

The final aspect of programming practice reinforced in this model version concerns good interface design. This includes use of colours (to highlight the epidemic front) and explicit consideration of the information most useful to the model user for presentation in plots and summary values.

6.11.2 NetLogo language

There are two types of primitives in the NetLogo language, commands and reporters. Commands change some aspect of the model, such as storing a new value in a variable. In contrast, reporters simply extract and provide information, such as recovering the stored variable value.

It is not a coincidence that this distinction exists for both primitives and procedures. Fundamentally, procedures and primitives are handled the same way by NetLogo; when NetLogo reaches one, it jumps to the piece of code with that name and receives instructions on what to do. The difference is that the code to run with a named primitive happens to be written by the NetLogo developers, while procedures are written by model developers (like you).

New NetLogo keywords explained in this model version:

- time: `reset-ticks`, `tick`
- space: `neighbors`
- code structures: `to-report`, `if`, `ifelse`, `ifelse-value`
- working with variable values: `set`, `of`
- working with agents: `ask`, `n-of`, `max-n-of`
- mathematics: `round`, `count`, `sum`, `!=`, `>`
- miscellaneous: `stop`

In addition, the primitive `myself` was used but not explained. A full understanding requires the concepts of agentsets and context, both of which are presented in the next chapter.

7

Model 3: Agents Making Decisions

Model 3 implements the behaviour decision process: people adopting (and dropping) protective behaviour. Once implemented, this version is a full agent-based model because it represents individuals enacting a process based on their situation and characteristics, so all the essential elements of the definition of an agent-based model are in place. That situation involves interaction with other individuals, which is important for most agent-based models in the social sciences.

From this point in the tutorial, there are fewer additional keywords introduced in each model version. Instead, this version mostly uses known keywords in new ways to work with new concepts. NetLogo is a rich language with many commands to implement specific actions, but much of model development is about using the main commands effectively to implement your design.

7.1 Model Design: Protective Behaviour Decisions

As for the epidemic process, a detailed specification of the behaviour decision process is required before considering how to implement that design in code. The broad design (see Section 4.3) states that people will behave protectively while the weighted combination of their attitude, behaviour of nearby people, and threat perception is above some threshold. Each of these behaviour inputs must be operationalised, as well as the calculation of the combination.

The first input is attitude, if a person is more positive about protective behaviour, then they would adopt such behaviour more easily. This is a personal characteristic that varies across individuals but is constant for each individual over time. While any scale can be used, it is convenient to operationalise attitude as a number in the interval [0,1].

The second input is the behaviour of nearby people. The most direct way to include this is to simply measure the proportion of nearby people that are currently enacting protecting behaviour. Initially at least, nearby will be operationalised as the same patch and neighbouring patches. As a proportion, the behaviour of nearby people will also be a number in the interval [0,1] like attitude. However, unlike attitude, it will change during the simulation as people adopt and drop their own protective behaviour, thereby changing the norm in their neighbourhood.

The final input is threat perception. This is related to incidence, the new infections this time step as proportion of population. The perception of threat should increase as the local infections increase and reduce as the epidemic passes. One way to implement this pattern is a discounted cumulative sum of local incidence, so that the influence of previous incidence levels decreases over time (that is, history fades). If δ is the discount, or amount of reduction of influence in a time step, then $(1 - \delta)$ is the incidence to be retained for threat perception. Such discounting continues indefinitely. For example, at time $t = 0$, threat would be initial incidence I_0, then $I_1 + (1-\delta)I_0$ at time $t = 1$, where I_j is the incidence at time j. At time $t = 2$, threat would be $I_2 + (1-\delta)I_1 + (1-\delta)^2 I_0$, which can also be written as $\sum_{j=0}^{2}(1-\delta)^{2-j}I_j$. More generally, the discounted incidence sum at any time t is given by $\sum_{j=0}^{t}(1-\delta)^{t-j}I_j$.

The relative importance of each of these factors in the decision can be controlled with weights, which should sum to 1. An individual's behaviour score is therefore given by:

$$B_i = \omega_A A_i + \omega_N N_{r1} + (1 - \omega_A - \omega_N)\sum_{j=0}^{t}(1-\delta)^{t-j} I_{t,r2}$$

where i specifies the individual, $r1$ and $r2$ denote the relevant regions for the perception of norms and threat respectively, t is the time step, A, N, and I represent attitude, norms and incidence, and ω_A and ω_N are the weights for attitude and norms. The behaviour score is compared to a threshold and behaviour is adopted or dropped accordingly.

In addition, people's protective behaviour must influence the epidemic process. The capacity of the epidemic to spread in a region (r) should reduce with increases in the proportion of people adopting protective behaviour (P_r) and the efficacy of that behaviour (E). Efficacy is the reduction in probability that a protected person is able to become infected or, if already infected, transmit the infection. For example, if 100% of the population have adopted protective behaviour and that behaviour is 40% efficacious, then infections should be 40% lower than they would be without such behaviour and, if 40% of the population have adopted protective behaviour and that behaviour is 100% efficacious, then infections should also be 40% lower than they would be without such behaviour. What happens if 60% of the population have adopted protective behaviour and that behaviour is 50% efficacious? This would reduce infections by 30% (calculated as 50% of 60%). Since β in the epidemic transmission equations represents that capacity, this parameter must be adjusted for local behaviour, with the adjustment given by:

$$\beta_r = \beta(1 - P_r E)$$

What is needed to implement this design? Try to identify the information (variables) and actions (procedures) required before moving on. Are the variables global or do they potentially have different values for some patches or turtle agents? Which procedures are required for initialisation (called by setup) or during the simulation (called by go)? All of this detail is part of the conversion from a design to a fully operationalised specification suitable for translating into code.

7.2 Context and Perspective

The basic representation in agent-based modelling takes the agent's perspective of its situation and acts accordingly. The need to take the perspective of different entities is incorporated in the NetLogo language and is formally referred to as context. As previously described (see Section 4.1.2), there are four types of model entities: observer, turtles, patches, and links. Each of those entities also supports a context.

Consider the code at Figure 7.1. Each version changes the colours of the patches along the diagonal to either red (left) or blue (right). You can test this by adding the procedure(s) to the code tab, then typing the name of the procedure in the Command Center.

The *diag-observer* procedure is written from the perspective of the observer, an external entity that can pass instructions to the model and, in some sense, stands in for the model developer or user. The first step is to `ask patches` to follow the instructions in the code block (delimited by [ ]). The `ask` instructs the simulation to randomly choose a patch, apply the code in the code block to that patch, then move to another randomly chosen patch until all patches have run the code. When interpreting (or writing) the code, you might find yourself imagining yourself as a patch and thinking about what happens to you. That is, all the code in the code block is from the perspective of the patch; you do not write the equivalent of 'look at the co-ordinates of the patch and then tell the patch to change colour if the co-ordinates match', instead the patches check their own `pxcor` and `pycor` and change colour if they are equal. In NetLogo terms, the opening bracket changes the context from the observer to the patch, the patch runs the code, and then the closing bracket changes it back.

Square brackets - context change

The square brackets [and] are used to change contexts in NetLogo (and for other purposes). For example, following a command by the observer to `create-turtles... [`, the context becomes turtles and the code delimited by the square brackets is run by each turtle as it is created.

The *diag-bypatch* procedure is also written from the perspective of the observer. It starts the same way, but then transfers control to a separate procedure. As before, the context changes with the square brackets. In this version, however, the different context occurs in a separate procedure. The do-patch procedure is written from the perspective of an individual patch and is run once for each patch (controlled by the `ask patches` in diag-bypatch).

The primitive `myself` can only be interpreted from an understanding of context. When one model entity instructs a second model entity to do something, that second entity may need information that is known to the original entity. But the switch of context with the `ask` statement xmeans that NetLogo is running code from the perspective of the second entity (while inside the square brackets). The `myself` primitive provides a link back to the first entity, the one issuing the `ask`, to access that information.

myself

`myself` is used inside an `ask` code block. It refers to the agent (or model entity) that issued the `ask`. See also `self`.

NetLogo keeps track of the context automatically. While most keywords can be used in any context (such as `ask` or `if`), others can only be used within certain contexts and attempting to use them from a different context will generate an error. These are marked by an icon in the NetLogo dictionary: patch has a checkerboard ( ), turtle agent has a dart or arrowhead ( ) and link has a chain link ( ). For example, `pxcor` can be used by patches and turtles.

```
to diag-observer
  ask patches
  [ if pxcor = pycor
    [ set pcolor red
    ]
  ]
end
```

```
to diag-bypatch
  ask patches [ do-patch ]
end

to do-patch
  if pxcor = pycor
  [ set pcolor blue
  ]
end
```

FIGURE 7.1: Running procedure diag-observer (left) or diag-bypatch (right) achieve the same outcome. The code on the left takes the perspective of the observer, instructing some patches to change colour to red. In the code on the right, the procedure diag-bypatch instructs each patch to run the do-patch procedure. The diag-bypatch procedure is observer context, and the do-patch procedure is patch context.

In the diag-observer procedure, the first line of `ask patches` can only be run from the observer context. NetLogo does not allow a patch (or turtle or link) to `ask` all the patches (or turtles or links) to do something. In the do-patch procedure, the first line refers to the patch specific primitive `pxcor`, signalling to NetLogo that the procedure is to be run by patches.

7.2.1 Updating threat, a patch procedure

Threat is calculated as a discounted sum of local incidence. While it is used by turtle agents in their decision making, it is the same value for all turtles on a patch. It is therefore more efficient (fewer calculations) to calculate at the patch level and simply have the turtles access the relevant threat value.

Remember from Section 7.1 that the threat (at patch r) at time t is given by a discounted cumulative sum of incidence (as a proportion of population) over time, with δ representing the discount. But calculating the sum would require keeping track of all the values of incidence since the start of the simulation.

A different way of thinking about this is that the new value of threat is given by adding together the new incidence (I_t) and the discounted value of the threat at the previous time step (T_{t-1}). This is because the threat at time step $t-1$ is the discounted incidence sum to that time step and captures all the historical incidence information.

$$\begin{aligned}
T_{t,r} &= \sum_{j=0}^{t}(1-\delta)^{t-j} I_{j,r} \\
&= (1-\delta)^0 I_{t,r} + (1-\delta)\sum_{j=0}^{t-1}(1-\delta)^{(t-1)-j} I_{j,r} \\
&= I_{t,r} + (1-\delta) T_{(t-1),r}
\end{aligned}$$

The update-threat procedure at Snippet 3a uses this relationship to calculate the new value of threat each tick. Snippet instructions continue to get briefer. The reminders to check syntax and save the model were dropped from Model 1 to Model 2 (hopefully you have remembered to do this each model element, and should continue to do so). Now, the snippets will include an instruction to call any new procedure, but the surrounding code from the *setup* or *go* procedure will not be displayed with the snippet code. Similarly, you are expected to know how to declare (create and make available) variables.

The procedure is a patch context procedure because the call to the procedure will be inserted within an `ask patches [ ]` block in the epidemic-spread procedure. Ordering is important: the spread-epidemic procedure must calculate incidence (along with epidemic state counts) before updating threat because incidence is required for the update.

Snippet 3a: Calculating threat

1. Create new global variable 'threat-discount'.
2. Add threat-discount to the initialise-globals procedure (with value 0.1).
3. Add 'threat' to the patches-own attributes.
4. Create update-threat procedure in the implementation procedures section.
5. Add the call to update-threat at the end of the last 'ask patches' in the 'spread-epidemic' procedure.

```
to update-threat          ; patch procedure
  set threat incidence / popn + (1 - threat-discount) * threat
end
```

Note that the variable *threat* is being used for both the assignment and within the calculation. The two values of 'threat' are different. This construction in NetLogo and many other programming languages is interpreted as: calculate the expression using the current value of the variable named 'threat' and then assign the results of that calculation back into the variable named 'threat'. There are three important concepts in this snippet that have been used already but not articulated:

- In programming languages, a structure like $x \leftarrow x+1$ is used to change the value of a variable (this is read as 'calculate the value of $x + 1$ and assign or allocate that value to the variable x). In many languages, the equals sign ($=$) is used as the assignment operator, which can be confusing if you read the expression $x = x + 1$ as an equation (that is, an assertion that the calculation on the left of the equals sign is the same value as the calculation on the right). In NetLogo, assignment is done with `set`, which is immediately followed by the name of the variable (always a single word, potentially with hyphens) to receive the value, and then the expression that calculates the new value.
- Variables remember their current value across time. That is, incrementing the internal clock with `tick` does not change or reset any variable values, the values are retained in the permanent storage set aside when the variables are created until they are explicitly changed.
- Accessing the current value of a variable is as simple as using the name of the variable in the expression.

This snippet makes no visible change, so the model remains as in Figure 6.12. However, a soundness check is necessary to ensure the code works as intended. Press *setup* and then open an inspect window for one of the patches where infections are present. Type 'go' into the Command Center to run the go procedure once. Look at the values of incidence and threat, then type 'go' again and look at the updated values. At tick 1, the value of threat should be incidence / popn. At tick 2, the value of threat should be $0.9 \times$ tick 1 threat + incidence / popn. More generally, inspect a patch outside of the epidemic area and note that the threat is zero until the epidemic is close then increases as the epidemic reaches the patch and decreases once the epidemic has passed.

7.3 Procedures with Arguments (Inputs)

So far, procedures have simply run the same set of commands each time, with the only variation in effect arising from the differences in the values of patch variables or environment. However, the full syntax of procedures allows inputs to be provided to the procedure. This is true for both command procedures (defined by `to`) and reporter procedures (defined by `to-report`).

If a procedure might be relevant to many different models, it is useful to have a general purpose procedure with arguments rather than fix values within the code as this makes it more flexible. Conceptually, this is how the NetLogo language is structured, each primitive is simply a procedure with arguments.

The other common use is reporter procedures that calculate something complicated that is needed by different parts of the code and changes as the situation changes. Conceptually, this is similar to mathematical functions that map numbers to some output. For example, the function $f(x) = x^2+3$ returns different values for different values of x. For $x = 2$, $f(x) = 7$, for $x = 5$, $f(x) = 28$ and so on.

In the same way, procedures can be written that include an input (referred to as an argument) in the processing of the code. The arguments are named immediately after the procedure name and included in square brackets for the procedure definition. When the procedure is called, the values for those arguments follow the procedure name without any special syntax.

Procedure argument naming

The arguments should have meaningful names. However, variables being passed could have very similar names, leading to confusion. It is therefore sensible to have some sort of naming convention that distinguishes between the agent variables and the procedure arguments. One such convention is to start the argument name with a special character such as a hash mark (#).

7.3.1 Assigning attitude, a procedure with arguments

In attitude surveys, a common response is 'agree', which results in a skewed distribution. To replicate this pattern, the attitudes will be randomly assigned from a triangular distribution with a mode of 0.8 (highest probability value).

NetLogo does not have a random number generator for the triangular distribution; instead we will draw a random number from the uniform distribution and apply a transformation. To convert a uniformly generated number U on the interval [0,1] to a triangular X on [0,1] with mode m, calculate:

$$X = \sqrt{Um} \qquad \text{for } U < m$$

$$X = 1 - \sqrt{(1-U)(1-m)} \qquad \text{for } U \geq m$$

Snippet 3b-1 creates a general procedure for generating random numbers from a triangular distribution with three parameters for the limits and the mode. It implements a more complicated version of the triangular conversion equation given for the interval [0,1]. The *triangular0to1* procedure calls that general procedure, passing the mode as an argument and setting the limits as 0 and 1. This latter procedure is then used to generate random attitude values for people agents in Snippet 3b-2.

Snippet 3b-1: Customised random numbers

1. Create random-triangular procedure in the utility procedures section.
2. Create triangular0to1 procedure in the utility procedures section.

```
to-report triangular0to1 [ #Mode ]
  report random-triangular #Mode 0 1
end

to-report random-triangular [ #Mode #Lower #Upper ]
  let uRand random-float 1
  let scale (#Mode - #Lower) / (#Upper - #Lower)
  report ifelse-value (uRand < scale)
  [ #Lower + sqrt(uRand * (#Upper - #Lower) * (#Mode - #Lower)) ]
  [ #Upper - sqrt((1 - uRand) * (#Upper - #Lower) * (#Upper - #Mode))]
end
```

random-float

The full syntax is `random-float n`. Generates a number anywhere in the continuous range from 0 to n (not just integers, but any number). The random number is drawn uniformly from the range, so any number is equally likely to be generated.

How can this code be checked? Random number generators can only be tested by drawing many numbers and examining the distribution of the values. However, there is no need for strict tests of randomness; that is the responsibility of the NetLogo developers in their construction of the

Snippet 3b-2: Assign (random) attitudes

1. Create a variable named 'attitude' for people.
2. Amend make-people procedure to add call to triangular0to1 to assign attitude.

Note: Shading is used for surrounding code that does not change and the ellipses (...) indicates that there is code that is not shown.

```
to make-people
    ...
    set size 0
    set attitude triangular0to1 0.8
    ...
end
```

random-float primitive. It is instead sufficient to examine whether the conversion to a triangular distribution is working as expected by examining the distribution.

Snippet 3b-3 constructs a histogram of the attitude values. It will populate after you press the *setup* button. What do you expect? Since you have assigned a mode of 0.8, you should see a triangular shape with the highest point at the value of 0.8 similar to Figure 7.2. Note that if you have only a single bar, you probably did not follow the instruction to set the interval (bar width) to 0.05.

Snippet 3b-3: Verify with histogram

1. Create a plot titled 'Attitude' with x-axis upper limit (X max) of 1 and the given code to create a histogram.
2. Use the pencil icon (at the end of the code entry line) to access the detailed settings and select bar mode with 0.05 as the interval.
3. Once the triangular distribution tests are complete, delete the plot (context menu on the plot).

```
histogram [attitude] of people
```

More thorough testing would vary the mode and see if the histogram adjusts accordingly. The most comprehensive testing would check the random-triangular procedure with different limits to make sure that the limits of 0 and 1 are not the only ones that work. Try code such as `set attitude random-triangular 3 2 5` (mode 3 with a range of 2 to 5) to assign attitude values; you may also need to set a higher value of 'X max' in the histogram settings to see these values. Remember to change the code back after testing.

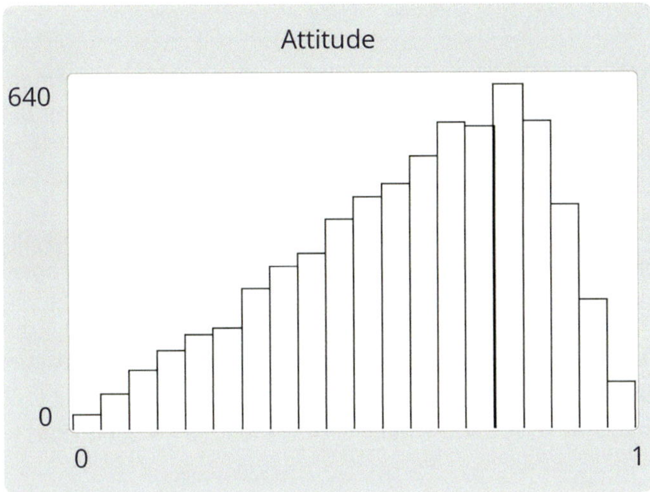

FIGURE 7.2: Screenshot of the histogram displaying the distribution of attitude values. Note the peak at 0.8 and the triangular shape.

Operationalising probability

The `random-float` reporter is particularly useful for implementing code that should be run with some probability. For example `if random-float 1 < 0.2` will be `true` with 20% probability.

7.4 Agentsets

An agentset is simply a group of patches or turtles or links but it cannot contain a mix of entity types. An agent can only appear once in an agentset; if an attempt is made to add an agent to an agentset in which it is already a member, nothing happens.

Whenever an agentset is accessed, for example with an `ask` command, the access is in a random order. That order is reset each time the agentset is accessed. This means that sequencing effects are automatically avoided; for example, no agent gets the advantage of always being the first to make a choice.

Agentsets provide much of the functionality of the NetLogo language. They support clear and concise coding of the fundamental representation in agent-based modelling, taking the perspective of each agent individually so that it can take some action.

So far, the tutorial model has extensively used the built in `patches` agentset, which is the group of all the patches. There are also equivalent agentsets for all `turtles` or all `links`. In addition, if breeds have been defined, then there is also an 'all members' type agentset for each breed. These agentsets are not only automatically available for any model, they are updated as agents are added (for example, with a `hatch` command) or removed (with a `die` command). Another important agentset built into the NetLogo language is `neighbors`.

patches

The `patches` agentset consists of all patches. It is automatically generated and updated (for example, if the size of the World is changed) and available to all model entities. See also `turtles` and `links`.

Agentsets can also be created explicitly with a `patch-set` or `turtle-set` or `link-set` command. This command instructs NetLogo to combine the agentsets that are specified into a single agentset. That is, these commands are the set operator 'union'.

The final method of creating agentsets is by selecting agents from an existing agentset. This has also been used already in the tutorial model with `n-of` and `max-n-of`. Another common subsetting command is `with` to select those agents that meet some condition, which will be used in Snippet 3e.

Agentsets can be assigned to a variable in exactly the same way as any other variable value, using a `set` or `let` command. But they can also be used in the code without being assigned to a variable. For example, in Snippet 1b-3, the instruction to assign a high population value to some patches was in the form: `ask n-of 3 patches [ set ... ]`. A set of 3 patches was created by the `n-of` statement and used by the `ask` command without ever being assigned to a variable.

Assigning agentsets to variables

If an agentset is to be used multiple times, it should be assigned to a variable. This is more efficient as the agentset only needs to be created once. While it is good practice to also store other information used multiple times, it is particularly important for agentsets as their creation can be computationally expensive if there are many agents in the model.

7.4.1 Agentset application

All the information needed by agents to decide whether to adopt protective behaviour now exists in the model. Snippet 3c-1 adds the procedure for people to make decisions. It uses all three types of agentsets discussed: built in, explicitly constructed, and subsets. All the new primitives included in the procedure concern agentsets.

The procedure sets the 'protect?' attribute owned by the turtle agents people. This variable has already been created in Snippet 1c-1, when demonstrating the `turtles-own` primitive, but the other variables needed to support the procedure must be created. Once you have entered and checked the code in the snippet, try and identify every agentset in the procedure.

There are several built in agentsets used in this snippet. `patches` and **`people`** are complete sets of patches and the turtle agents 'people' breed, respectively. The agentset `self` has only a single member, the agent running the code. The final built in agentset used is `neighbors`, introduced in Section 6.9.2.

Snippet 3c-1: Decide behaviour

1. Add two variables to the **patches-own** list: prop-protect, max-threat
2. Create three sliders for weights and threshold with range 0 to 1 and increment 0.05: wt-attitude (value 0.4), wt-norms (value 0.2), threshold (value 0.35)
3. Create decide-behaviour procedure (in the implementation procedures section)
4. Add call to decide-behaviour as the last line in the go procedure, before **tick**

```
to decide-behaviour
  ask patches
  [ let visible (patch-set self neighbors)
    set prop-protect (count (people-on visible) with [protect?]) /
        (count people-on visible)
    set max-threat max [threat] of visible
  ]
  ask people
  [ let decision-value wt-attitude * attitude + wt-norms *
         prop-protect + (1 - wt-attitude - wt-norms) * max-threat
    set protect? ifelse-value (decision-value >= threshold)
         [TRUE] [FALSE]
  ]
end
```

self

`self` is the agentset comprised of one agent, the one from whose perspective the code is being implemented. See also `myself`.

The `patch-set` primitive is used to explicitly construct a 3×3 group of patches as the union of the patch at the centre (with `self`) and its eight neighbours (with `neighbors`). The union patchset is stored in the local variable named 'visible', which is temporarily created for each patch in turn. The `let` command creates that variable and assigns the agentset as its value. The 'visible' agentset is then used to calculate values for the prop-protect and max-threat variables for the patch, and discarded at the end of the code block.

The calculation of max-threat demonstrates two agentset keywords that have already been used. The patchset named 'visible' has nine members; the values of the variable 'threat' are extracted[1] for each of those patches using `of` and the maximum of those nine values is found with `max`. This is the same construction as finding the `sum` of popn-I values to decide whether the epidemic has ended (at Snippet 2e-1), and for calculating the maximum population over all patches (in Snippet 1b-3). The difference is that `of` is here being used for a constructed agentset rather than a built-in one as used previously.

[1]Technically, these are stored in a list, discussed in Section 12.2.

patch-set

The full syntax is (`patch-set` **`input-patchset-A input-patchset-B input-patchset-C ...`**). Note that the brackets are essential. This is a union operator to combine multiple specified patchsets into one patchset. Even if a patch appears in more that one of the input patchsets, it will only appear once in the combined patchset. See also `turtle-set` and `link-set`.

The other agentsets are more subtle. The new primitive **`people-on`** is a breed specific version of `turtles-on`, which constructs an agentset comprised of all the turtles that are located on the specified patchset, in this case the 3×3 group of patches named 'visible'. So the new agentset is all the people that are nearby, either on the same patch or one of the neighbouring patches. A related agentset is constructed using `with` to create a subset of these people, selecting only those with the variable named 'protect?' set to `true`. The number of people in each of these agentsets is calculated with the primitive `count`, and the quotient of these counts gives the proportion of nearby people who are currently adhering to protective behaviour. Neither of the agentsets involved are ever stored in a named variable, they are implicit and exist only during that calculation.

turtles-on

The full syntax is `turtles-on` **`<some-patchset>`**. Creates an agentset of all the turtles that are located on the patches in the specified patchset. The turtles can be all turtle agents, or just those of a specific breed.

with

The reporter `with` subsets an agentset, creating a new agentset that only contains the agents that meet the specified condition.

7.5 Decisions with Thresholds

The other part of Snippet 3c-1 combines each person's attitude with the patch information about perceived threat and norms (the variables 'max-threat' and 'prop-protect' respectively) to decide whether the person adopts or drops protective behaviour. The weighted sum of the three inputs (with weights set by the sliders) is compared to the behaviour threshold value (also set by slider).

Snippet 3c-1 sets the value of 'protect?' to `true` or `false` depending on the results of a conditional test, which also returns `true` or `false`. The truth value has been explicitly assigned for clarity, but there is a shorter equivalent (see Figure 7.3) that takes advantage of the fact that truth values are consistent; it doesn't matter whether it is a boolean (logical) variable or a test, `true` is `true`.

By now, you should be considering what soundness check to use for this code. One option is to run the model and stop it when the epidemic is very active. You can then simply count the people currently protected by typing code into the Command Center. Suitable code would be

```
set protect? ifelse-value (decision-value >= threshold) [TRUE] [FALSE]
set protect? decision-value >= threshold
```

FIGURE 7.3: These two lines of code are equivalent. The first explicitly sets the truth value of the variable based on whether the test is true. The second simply assigns the result of the comparison and is shorter, but more difficult for a human reader.

`count people with [protect?]`. You could also `inspect` some people and the patch where they are located to see the values of the various decision inputs and the protection state of the person. Or you could add a monitor to the interface to count the protected people and let the model run, deleting the monitor when you are confident the model is operating correctly.

An easier way to check whether the protective behaviour process is working as expected is to construct a plot of the proportion of the population who are protected. This is also useful information for users, and the plot to add is described at Snippet 3c-2. With a plot, it is easy to test the effect of different threshold values; with a low threshold almost every person should be protected, and none would be protected if the threshold was set very high. However, such a plot does not negate the need to inspect some people to ensure the detailed calculations are correct.

Snippet 3c-2: Summarise protected people

1. Create a plot titled 'Proportion protected' (with Y max of 1 as it is a proportion)

```
plot count people with [protect?] / count people ; code for pen
```

Once the plot is created, press the *setup* and *do it!* buttons. The model should be similar to the output shown in Figure 7.4. Notice that the interface widgets are grouped so that epidemic controls are together, as are protective behaviour settings.

Interface design

Organise the interface to make the model easy to use and understand. If there are only a small number of controls, ensure related widgets are close to each other. For larger models, it may be appropriate to have the main controls and output in one area and more advanced settings that are used less frequently in another area.

7.6 Finishing off the Core ABM

The processes of the ABM are now in place; an epidemic spreads and people react to the threat and the decisions of others by adopting protective behaviour (or not). However, that behaviour is not yet influencing the epidemic. From the model design at Section 7.1, a patch's capacity

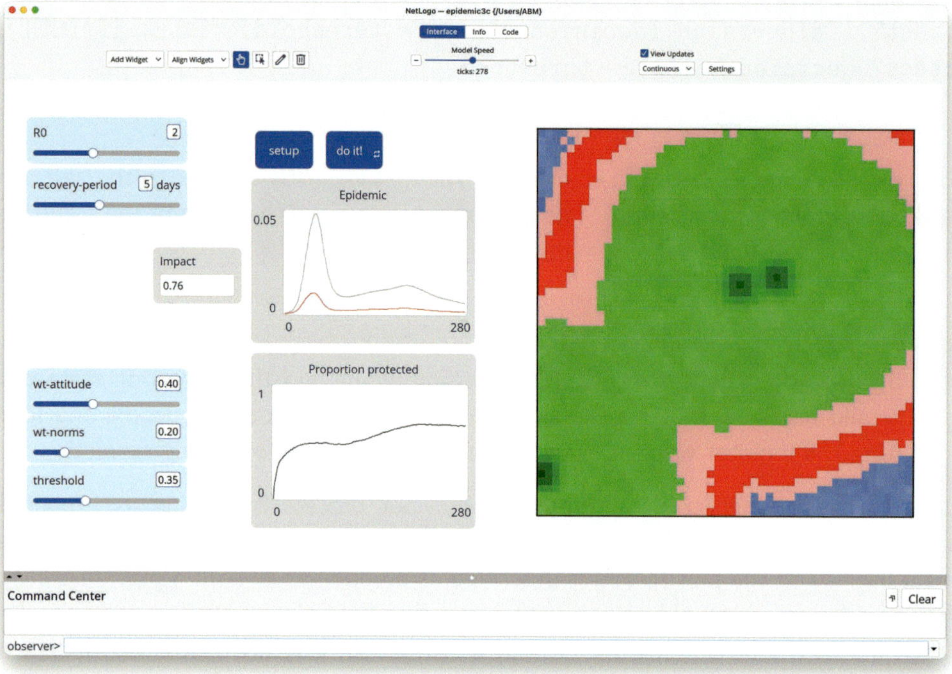

FIGURE 7.4: Screenshot of the interface after completing version 3c of the model. Approximately half the population quickly adopts protective behaviour, with protection levels starting to decline as the epidemic is dying out.

to transmit the epidemic must be adjusted for the level of protective behaviour and its efficacy using the equation: $\beta_r = \beta\,(1 - P_r E)$. This is implemented in Snippet 3d.

The new agentset primitive, `people-here`, is a breed specific version of `turtles-here`, which creates an agentset of all the turtles at the patch issuing the instruction. This enforces the local effect of protective behaviour, as only people on the relevant patch influence the epidemic process.

turtles-here

`turtles-here` creates an agentset of all the turtles that are located on the instructing patch, or the patch where the instructing turtle is located. The turtles can be all turtle agents, or just those of a specific breed.

It is straightforward to test this snippet, by simply adjusting the efficacy slider and seeing the effect on epidemic impact. Experiment with different levels; an example where the behaviour stopped the epidemic from fully taking off is shown at Figure 7.5. Remember also that the effect of protective behaviour on the epidemic depends on both the efficacy and how much of the population adopt protective behaviour (which can be controlled with the threshold).

Protective behaviour comes in many forms including vaccination, increased hand washing, or reducing social contact. The model should be able to represent any of these behaviours. They vary in efficacy, which can be adjusted with the slider that has just been added. However, they also vary in permanency. Once a person is vaccinated, they can't abandon the behaviour.

Snippet 3d: Behaviour influences epidemic

1. Create slider for 'efficacy' (0 to 1 by 0.01, value 0.8)
2. Amend spread-epidemic procedure.

Note: Shading is used for surrounding code that does not change and the ellipses (...) indicates that there is code that is not shown.

```
to spread-epidemic
  ...
  ask patches
  [ ; calculate the number of new infections generated by the patch
    let beta (R0 / recovery-period)  * (1 - efficacy *
          (count people-here with [protect?] / count people-here))
    let new-cases popn-I * beta * popn-S / popn
    ...
end
```

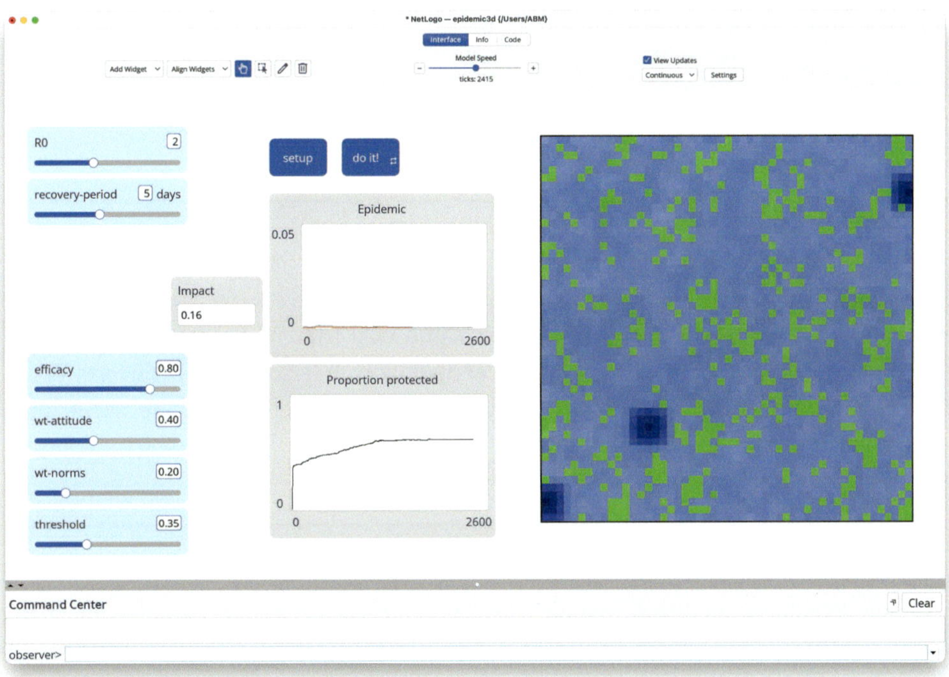

FIGURE 7.5: Screenshot of the interface after completing version 3d of the model. The effect of protected behaviour is visible, with much of the World coloured blue at the end of the epidemic, indicating that less than half the population were ever infected.

The final step of model version 3 is to add a user control to set whether the protective behaviour is reversible. This is implemented with a new type of input widget called a switch. A switch is used for boolean variables, it is either on (true) or off (false). As with other widgets, it is added directly to the interface and its value assigned to a global variable simply by entering an appropriate variable name.

Snippet 3e: Some decisions are final

1. Create a switch for the variable 'reversible?'
2. Amend decide-behaviour procedure.

Note: Shading is used for surrounding code that does not change and the ellipses (...) indicates that there is code that is not shown.

```
to decide-behaviour
  ...
  ask people
  [ let decision-value wt-attitude * attitude + wt-norms *
       prop-protect + (1 - wt-attitude - wt-norms) * max-threat
    ifelse reversible?
    [ set protect? ifelse-value (decision-value >= threshold)
         [TRUE] [FALSE] ]
    [ if not protect? and (decision-value >= threshold)
         [ set protect? TRUE ] ]
  ]
end
```

Logical operators

Logical operators manipulate truth values in the same way that mathematical operators (such as +) manipulate numbers. Common logical operators are: not which converts true to false and false to true; and which is true only if all input values are true; and or which is true if any input value is true.

Working with logical variables

Use logical operators and condition testing when working with boolean variables because NetLogo will expect a logical value (true or false) and report an error if it encounters some other value. There are some subtle distinctions; for example, if [test?] will report an error if 'test?' has value 0 because if expects to encounter something that returns true or false. However, if [test? = true] will appear to run correctly; the test will always return false because 0 is never equal to true but the code will run since a value of false is returned.

What soundness check can you think of, what behaviour do you expect when using the switch? Change the efficacy slider to 0 and run the model twice, once with the switch on, and once with the switch off. Because efficacy is set to 0, the level of protective behaviour has no effect on the epidemic impact and there should be little difference in model results. However, in the runs with the switch set to not reversible, the proportion protected should increase (or remain stable), even when the epidemic has passed. This can be seen in Figure 7.6.

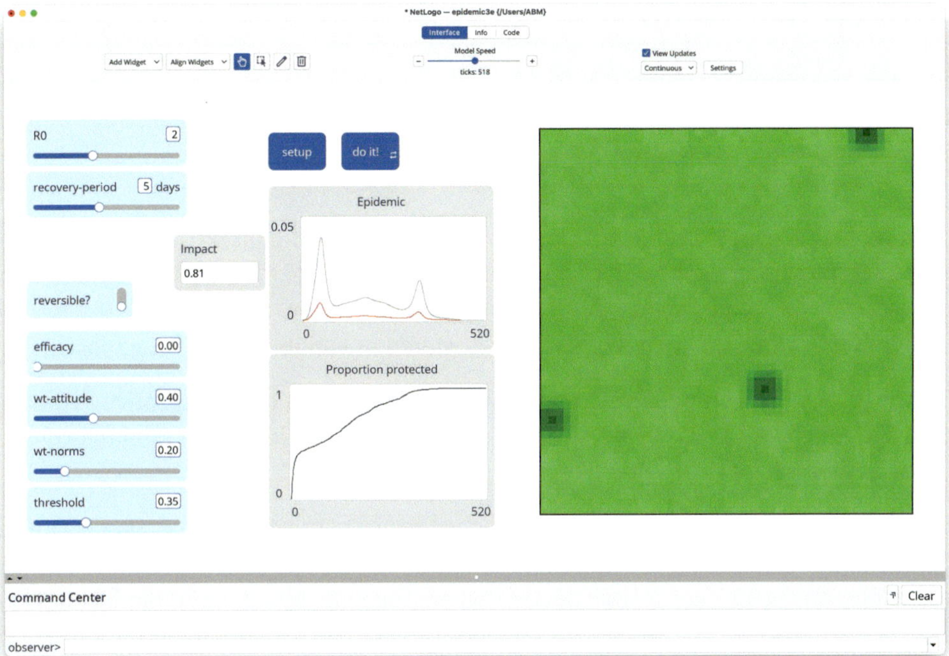

FIGURE 7.6: Screenshot of the interface after completing version 3e of the model. The effect of irreversible behaviour is displayed in the proportion protected plot, which increases throughout the epidemic and does not decline once the epidemic is over.

7.7 Tutorial Progress Check

Version 3 of the model is now complete. Referring back to the definition (see Section 1.4), it is now an agent-based model, as it:

- is a computer program. . . : yes, it is written in the NetLogo language and runs on the computer;
- that represents individual entities. . . : yes, the turtle breed 'people' are individuals;
- taking actions. . . : yes, the simulated people adopt and abandon protective behaviour;
- in accordance with their own characteristics, resources, beliefs. . . : yes, the personal attribute of attitude is a component in their choice of action;
- and perception of their social and physical environment.: yes, the simulated people are able to perceive the proportion of people nearby that are behaving protectively and the epidemic threat, both of which influence the choice of action.

Having now directly implemented these features in the tutorial model, the meaning of each aspect of the definition should be reasonably clear. Agent-based models are able to represent complex systems because of these features, as they allow interaction and its effects to be represented. The behaviour of the epidemic is driven by the interactions between the epidemic and protective behaviour processes. This was immediately apparent when the processes were connected in Snippet 3d and the epidemic was throttled by effective protection.

Keep this version

This version is the simplest complete ABM in the tutorial. Keep a copy (even if not saving the other intermediate versions) as it is used in Chapter 13 to demonstrate batch simulation.

Two important concepts were introduced in model 3: agentsets and context. Agentsets are fundamental to the NetLogo approach to coding an agent-based model. They allow concise identification and selection of relevant agents, both overtly or more subtly, underpinning the flexibility of NetLogo. Agentsets can be combined or subsetted, stored in variables, and issued instructions.

Each time NetLogo accesses an agentset, the context changes and the perspective is taken of each individual agent in turn. The order in which NetLogo iterates through the agents is random, and is reordered each access.

When developing code, it is often easiest to imagine yourself as an agent (such as a turtle) and think about what information you would need and what you would do to perform the required task; that is, adopt an agent-centric perspective when thinking about your model. It is also worthwhile to consider whether the same logic applies for all agents, particularly the ones taking the first turn, a middle turn, and the last turn.

In addition, two previously introduced concepts were extended: procedures and working with truth values. Procedures are used to modularise the code and either instruct agents to perform some task (command) or provide some information (reporter). The more advanced feature introduced is the use of arguments, inputs that the procedure can use for more flexibility. The tutorial model used this capacity to construct a procedure that provides random numbers from a triangular distribution with the limits and mode of that distribution passed as arguments. This potential for arguments emphasises the equivalence between user written procedures and the primitives built in to the NetLogo language. Primitives are simply (efficiently written and robust) procedures with arguments.

Truth values (that is `true` and `false`) are included in models in different ways. They are used as status flags, such as the variable 'protect?'. They are also generated as the outcome of comparisons and other tests. Such tests are overt in the construction of `ifelse` and similar constructions, but are also fundamental to agentsets constructed with the `with` filter. These truth values are their own type of value. That is, the truth value `true` is not the same as the text string "true", nor is it stored as a number (such as 1, as occurs in some languages) with a specific indicator on the variable type. Logical operators should be used to work with truth values, in the same way that mathematical operators are the most appropriate for numbers (or string operators for text).

7.7.1 Programming practices

Version 3 of the model is now at the point that complex behaviour is being simulated. Each step has been only a small addition that is tested before moving on. Such iterative construction and testing is important when programming in any language, but it is essential for NetLogo. This is because the normal methods available to debug programs in other languages, such as stepping through the code and examining variable values, are much harder to do in NetLogo. Furthermore, logical errors are very difficult to detect (in contrast to syntax errors, which immediately generate an error message).

Each version of the model has been presented with less detailed instructions. At this point you should be comfortable with adding widgets to the interface, and automatically syntax checking and saving your model. You should also be starting to recognise the syntax of common primitives such as `ask` (which must be followed by an agentset and a command block) and `set` (which must be followed by a variable name and the value to be assigned).

If you are not already doing so, you should try and develop the habit of thinking about algorithm design, operationalisation, and implementation in code for every additional section of code. Algorithm design means working out each step that must be programmed to instruct the computer how to implement the model. This also helps you think about what tests you need to check that the code works as expected. The tutorial presents the full design of the model and also some tests so you are sure the design is implemented as intended. But these instructions are not for you to blindly follow, you need to think about why you are doing what you are doing.

Another programming practice built into the tutorial is interface design. The instructions provide details of what widgets to include on the interface, and the screenshots display one option for placement. What to include and where to place it are part of the model design. The model must be accessible for users. Parameters that need to be adjusted should be available, but too many widgets adds to the clutter and makes it difficult to find the correct one. The world displays the current status of the model, but colours and other features can be used to emphasise or hide particular aspects of that status. Finally, plots and other output should be focused on the information needs of the user.

This does not mean, however, that the interface cannot include widgets for the convenience of the model developer, they should just be removed from the released version of the model. In fact, it can be very useful to have an area of the interface devoted to diagnostic information to make sure that changes do not introduce errors in already tested code. The attitude distribution plot and maximum population monitors are both examples of such diagnostic widgets.

7.7.2 NetLogo language

Most of the new primitives introduced in this chapter concern the construction of agentsets. New NetLogo keywords explained in this model version:

- agentsets: `patches`, `turtles`, `patch-set`, `turtles-on`, `turtles-here`, `with`
- context: `self`, `myself`
- mathematics: `random-float`
- logic: `not`, `and`

Questions

1. What features of NetLogo make it particularly suitable for agent-based modelling?
2. Using the NetLogo Dictionary (help menu), look up three important keywords that are used in the tutorial model. If you don't fully understand the explanations, revise the section of the tutorial where the keyword was introduced.
3. What errors have you made while implementing the tutorial model? What made you realise you had made an error and how did you fix it?

8

Model 4: Representing Relationships

While version 3 of the model is a complete agent-based model, it does not fully implement the original design. In particular, Section 4.3 states that people communicate with their friends about threat. This is the element to be added in version 4, which introduces the final type of model entity, that of links. Links represent relationships between pairs of turtles.

8.1 Introducing Links

A network is an abstract mathematical way to represent relationships between entities, comprised of nodes for entities and edges for relationships (Newman, 2010). Nodes can represent people or firms or countries or many other real world entities. Relationships are equally broad, such as friendship or shared board membership or international trade. In NetLogo, network nodes are equivalent to turtle agents and links are used to represent relationships or network edges.

Links are a new type of NetLogo model entity (confusingly referred to as types of agents in the official documentation). The four entities are observer, turtle, patch, and link. The observer is external to the model, patches are used to represent space, turtles to represent agents (as meant in this tutorial), and links to represent relationships between pairs of turtle agents.

Much of the apparatus that you have already been using for patches and turtles is easily adapted to links. Some primitives work in exactly the same way, including `ask` and `set`. Others have equivalents, for example `link-neighbors` creates the agentset of turtles at the other end of the links connected to the asking turtle(s), which is the link version of the spatial `neighbors`. Therefore, while there are many primitives introduced in this chapter, they are mostly related to previously seen primitives rather than new programming concepts.

Like turtles, different breeds can be defined for links. Each breed has variables, defined with a `links-own` statement (or its breed equivalent). These variables store attributes of the relationship, not attributes of the turtles at each end of the link. For example, suitable variables for friendship links would include the number of times per month that the two people (represented by two turtles) contact each other or when they first met.

8.2 Model Design: Influential Friends

In order to communicate with friends, people must have friends. Friendship is a relationship between people, not an attribute of a particular person or a feature of the environment, and will be modelled with links. While not particularly realistic, the model will randomly allocate the same number of friends for each person. Furthermore, the friendships will be undirected, which means that if person A considers person B a friend, then person B also considers person A a friend.

The other design decision is how to interpret communication about threat. Currently, threat is one of the three inputs to the protective behaviour decision. That threat is based solely on the local epidemic incidence and is the same for all people on a patch (and consequently calculated by patches). Instead, threat is to be the maximum of the threat across all a person's friends. That is, if a person is in a location where the epidemic is active and therefore perceives a high threat level, that heightened threat is communicated to all their friends and the friends adjust their perceived threat accordingly.

8.3 Creating a Social Network

Creating a network with a fixed number of edges per node is not straightforward. If you simply ask every node to create (say) two edges with randomly selected other nodes, then some nodes may be selected more often than others. Instead, each node must be aware of the target number of edges and only accept selection if they have not yet reached that target. This algorithm is implemented in Snippet 4a.

The primitive `my-links` is the agentset of links connected to the specific turtle. For the first person, the count of this agentset is 0, so the local variable 'needed' is set to the value of the num-friends slider. For later people, some links may already have been created and needed is accordingly lower. If any links need to be created for that person (that is, `if needed > 0` is `TRUE`), the code block to create links is run.

my-links

The agentset of links connected to a specific turtle is generated with the `my-links` primitive. See also `link-neighbors`, which is the agentset of turtles at the other end of those links.

That code block selects all the people who still have capacity for further links, and excludes the person looking for friends (with the `other` primitive), assigning that agentset of people to the variable named candidates. But what if there's not enough people available to connect to? NetLogo will report an error if it tries to select too many agents (for example, if you are asking for 5 random turtles from a set with only 3). So the value for the `n-of` needs to take the minimum of the number to be selected and the number available.

Snippet 4a: Create a social network

1. Add slider for 'num-friends' (0 to 5 by 1, value 2)
2. Create make-network procedure
3. Add call to make-network procedure in setup procedure (think about where, you need people to link)

```
to make-network
  ask people
  [ let needed num-friends - count my-links
    if needed > 0
    ; only create links with people who have capacity
    [ let candidates other people with
                [ count my-links < num-friends ]
      create-links-with n-of  ; not more links than targets
                min (list needed count candidates) candidates
      [ hide-link]
    ]
  ]
end
```

other

The primitive `other` excludes the turtle or patch that is constructing an agentset from the agentset being constructed, even though it satisfies the conditions for inclusion.

The `min` primitive takes the minimum of a list. The tutorial has previously used `max` to find the maximum popn value. In that case, a list was automatically created by the `of` reporter requesting the values of a variable from all patches. Here, however, the list must be explicitly created with the values that are to be its items. For example, `(list 1 2 3)` would create a list of three items, the numbers 1, 2, and 3. Lists are discussed at Section 12.2.

The smaller of the number of links needed and the number of available turtles to link to is passed to the `n-of`. That number of people are randomly selected from the agentset stored in the variable 'candidates' to generate an unnamed agentset.

The primitive `create-links-with` instructs the person to create an undirected link with each of the people in the unnamed agentset. Like the turtle creation commands discussed in Section 5.8, the code block that immediately follows is run for each link as it is created. In this snippet, that code block hides the link so it is not shown in the world.

create-links-with

The full syntax is **create-links-with <agentset> [ ]**. Instructs a turtle to create an undirected link with every turtle in the specified agentset. Any code within the square brackets is run for each link as it is created. See also `create-links-from` and `create-links-to` for directed link versions. Similar primitives are also available that specify the breed of the links to be created.

Since the links are hidden, they are not available for the popup menu. Instead, type `inspect one-of links` into the Command Center to open the inspect window for a random link. You can see the built in variables for a link. A specific link is identified by the `who` numbers of the two turtles it joins (top of inspect window).

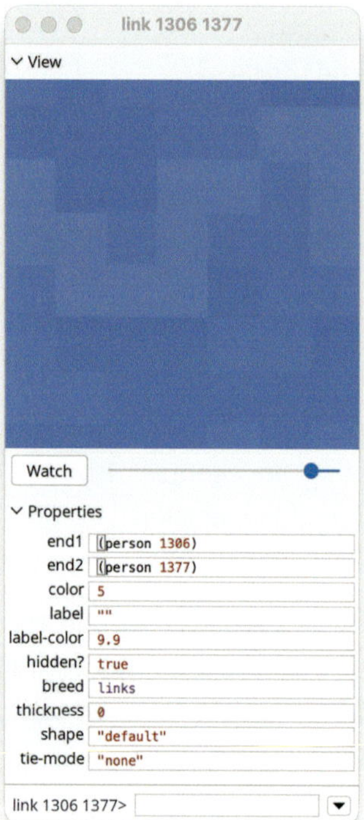

FIGURE 8.1: Inspect window for link.

No screenshot is presented as it is identical to Figure 7.6. The most obvious effect is that initialisation (the code run with the *setup* button) takes much longer to run. This is because each link being created requires NetLogo to construct an agentset by checking every person to count its current number of friends. You may also see the world apparently flickering as the new links briefly appear before being hidden.

Knowing that the code creates links is not, however, a check that it creates the correct links. With a little thought, however, you may realise that setting the slider for num-friends to 2 gives you access to a simple test. If each link connects two people and each person has two friends, then the number of people and the number of links should be identical. You can test this by typing 'count people' into the Command Center and then 'count links'. It may be that the number of links is slightly lower; if the last person asked does not already have any friends, then the only candidate is itself and links cannot be made from a turtle to the same turtle.

Finding the person agentset and the linkset to be the same size is not a conclusive test because links created at random will still give rise to the correct number of links. What you want to check is that all people have exactly two friends (or equivalently, exactly two links), with the possible exception of the failed self link attempts. One possibility is a histogram of `count [my-links] of people`. But it is easier to simply find the maximum of the number of links. If the maximum is 2, then all people must have two friends. The appropriate code is `max [count my-links] of people`.

hide-link

The command `hide-link` makes a link invisible in the world view. It is reversed with `show-link`. These are the link equivalents to `hide-turtle` and `show-turtle`.

8.4 Using Links

Using the links to obtain information from friends introduces a new NetLogo agentset, that of `link-neighbors`. This is the links equivalent of `neighbors`, generating an agentset of turtles that are socially connected to the instructing turtle, instead of patches that are spatially next to the instructing patch. The reporter `of` accesses information from agentsets, as seen previously (Snippet 2e-1). Snippet 4b uses the same primitive to access information about threat values where friends are located. No screenshot is displayed as the changes have no visible effect.

Snippet 4b: Communicating along links

1. Amend decide-behaviour procedure.

Note: Shading is used for surrounding code that does not change and the ellipses (...) indicates that there is code that is not shown.

```
to decide-behaviour
  ...
  ask people
  [ let friend-threat ifelse-value any? link-neighbors
       [ max [threat] of link-neighbors ] [ 0 ]
    let mythreat max (list max-threat friend-threat)
    let decision-value wt-attitude * attitude + wt-norms *
         prop-protect + (1 - wt-attitude - wt-norms) * mythreat
    ifelse reversible?
    ...
  ]
end
```

link-neigbors

`link-neighbors` is the agentset of turtles at the other end of the links connected to the specified turtle. See also `my-links`, which is the agentset of links representing those connections.

NetLogo will fail if it is instructed to run code on an empty agentset. The user may set number of friends to 0. Therefore the calculation of the maximum of the friends' threat values is preceded by a protective test as to whether there are friends to calculate over. If not, the friends' threat value is set to 0. This protection also deals with the case where a non-zero number of friends is selected but there are no potential friends available for the final person to have edges created.

any?

`any?` is a logical operator that reports `true` if the specified agentset has at least one member, and `false` if is empty.

Error protection on empty agentsets

If it is possible that an agentset is empty, it is good practice to precede any `ask` on that agentset with a test using `if any?`.

This snippet also uses the explicit `list` constructor. Applying `of` to the agentset of `link-neighbors` implicitly creates a list of the threat variable values so that the maximum of the threat values of friends can be calculated. The explicit `list` is used to create a list with two members, that maximum and the previously calculated maximum threat from the directly observed patches. Once these two values have been placed in a list, the `max` operator chooses the larger to be the relevant perceived threat. The snippet requires only minor amendments to the previous code because the revised maximum is inserted into the behaviour decision by simply replacing the maximum already present that refers only to the observed patches.

list

The full syntax is (`list` <entry0> <entry1> <entry2> `...`) to create a list with the items specified as its contents, in the order specified. List entries can be numbers, text strings, other lists, agentsets, or some combination of these data types.

There is no obvious way to test this code as the threat combining the sources is stored only as a local variable and therefore cannot be queried while the model is stopped. However, there are primitives that print information to the Command Center and these can be used to report relevant values at specific points in the code. Such primitives are commonly used to construct a basic test.

Save the model so that you can revert to the saved version after testing. Make the changes at Figure 8.2, but do not save the model. This code modifies the decide-behaviour procedure to provide information about variable contents. The primitives `type` and `print` are used to report the (labelled) threat inputs and then the threat value used. There are also equivalent commands to send information to an output widget on the interface or to a file. Note that the `ask people` has been modified to `ask n-of 10 people` to limit the amount of information sent each tick.

Press the *setup* button and type *go* into the Command Center (to run the go procedure once only). Remember (from Section 6.3.3) that you can expand and collapse the reporting area with the double ended arrow in the top right corner of the Command Center. As only 10 people are being queried, you may need to enter *go* more than once to find a person whose friends have higher threat values and therefore see a change in the person's threat value. Once you have checked the code, close the model without saving (assuming you haven't saved these changes) or reverse the edits.

```
to decide-behaviour
  ask patches
  [ ...
  ]
  ask n-of 10 people
  [ ...
    ... [ 0 ]
    type "patch threat is: " print max-threat
    type "friend threat is: " print friend-threat
    let mythreat max (list max-threat friend-threat)
    type "threat is: " print mythreat
    let decision-value...
  ]
end
```

FIGURE 8.2: Modifications to use output printed to the Command Center to confirm code is implemented correctly.

print, type

`print` and `type` both instruct NetLogo to provide information to the Command Center. The difference is that `type` sends only the information asked, while `print` sends the information and ends the line. See also `show` and `write`, which also send information to the Command Centre but additionally includes the identifier for the model entity.

8.4.1 Measure the effect of extended threat perception

The final step in this version is to report the effect of friends on behaviour. Each friend potentially increases perceived threat and therefore triggers adoption of protective behaviour earlier or delays cancellation. Therefore, more friends could be expected to increase protective behaviour overall.

Potential ways to measure this include the average proportion protected over the duration of the run or the maximum adoption level achieved. Both of these summarise the whole of the simulation, not simply measure the current status. Snippet 4c implements tracking the maximum proportion protected achieved during the run and when that maximum first occurs. Each time step it calculates the current proportion of people with protective behaviour. If that value is higher than the stored maximum, that maximum is replaced and the current time is also stored.

Note that there is a complication arising from the effect of that protective behaviour; with the epidemic restricted, threat is lower, confounding the effect of the friends. To avoid this, any simulations to test the hypothesis should be run with efficacy set to zero.

This snippet demonstrates the use of `ticks` to obtain the current time during the simulation (as compared to `tick` to advance the tick counter). Each tick, the current value of protected

Snippet 4c: Measure protection

1. Create two global variables: max-protect and when-max-protect.
2. Initialise both new variables to 0.
3. Create the update-globals procedure.
4. Add call to update-globals procedure at end of go procedure (before **tick**).
5. Add monitors to the interface for each new variable.

```
to update-globals
  let protected count people with [protect?] / count people
  if protected > max-protect
  [ set max-protect protected
    set when-max-protect ticks
  ]
end
```

proportion is compared to the stored maximum already achieved. If the new value is higher, both the maximum and the time at which that maximum is achieved are updated.

ticks

`ticks` queries the tick counter and reports the current time.

Once implemented, this snippet completes version 4 of the model, displayed at Figure 8.3. No explicit test is required for this code, because it is easy to check whether the new monitors are correct by comparing their reported values with the plot of proportion protected. Note that hovering over the plot (placing the mouse pointer near the line) will reveal the plotted values.

8.5 Tutorial Progress Check

Version 4 of the model is now complete. The purpose of this version was to introduce links, the final NetLogo model entity type. Links can have breeds and attributes; neither are included in the tutorial, but the concepts work in the same way as for turtles. Many primitives for patches and turtles you have already seen in previous versions work in exactly the same way for links, such as `ask` and `set`. Other turtle or patch primitives have equivalents for links with appropriately modified names, such as `link-neighbors` and `create-links-with`. Finally, there are primitives such as `my-links` that have no clear equivalent for other entities.

Programming practices continue to be reinforced. These include detailed design before coding, modularisation of code, minimum feasible change, syntax, and logic testing for each change. A new way to test was introduced, printing variable values to describe changes that occur within a time step.

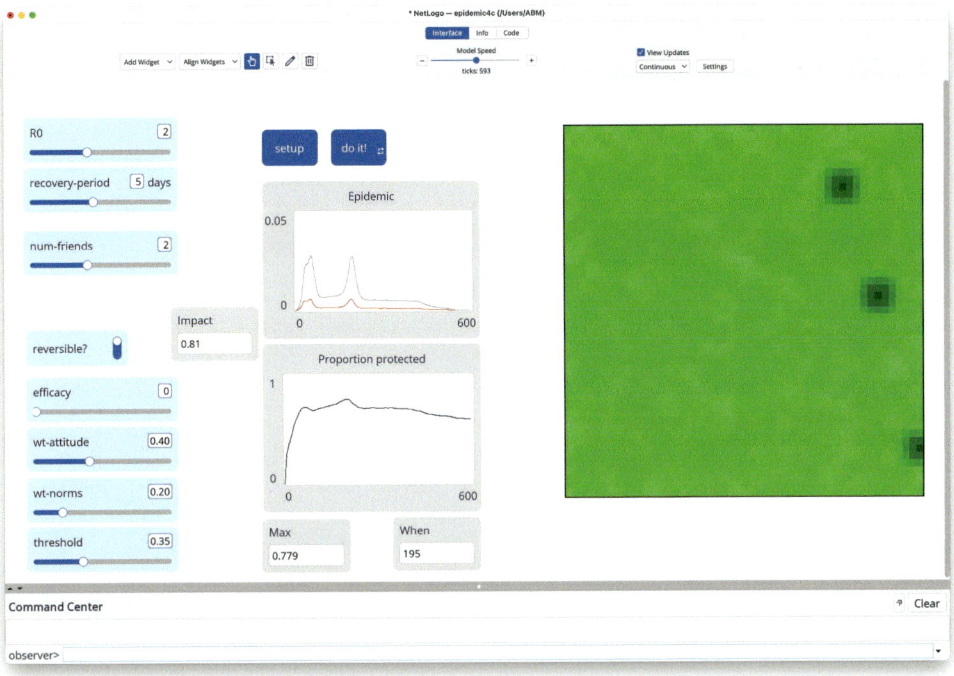

FIGURE 8.3: Screenshot of the interface after completing version 4c of the model. Note that the values in the two monitors concerning protective behaviour coincide with the peak in the protective behaviour plot.

8.5.1 NetLogo language

Most of the new primitives introduced in this chapter concern the construction and application of links. New NetLogo keywords explained in this model version:

- links: `create-links-with`, `my-links`, `link-neighbors`, `hide-link`
- agentsets: `other`, `any?`
- other: `list`, `print`, `type`, `ticks`

9

Models 5-8: Enhancements

The core NetLogo concepts and language have been presented in model versions 1 to 4. Version 1 established the model entities to represent agents and physical environment: turtles and patches. Version 2 introduced time and space, and implemented the process of epidemic spread in the patches. Version 3 implemented agent decisions, the process of adopting and dropping protective behaviour, relying extensively on the key concept of agentsets. Finally, version 4 introduced links to represent relationships.

The rest of the main tutorial enhances the existing model. Each enhancement is independent. They are intended to stimulate ideas for what can be done with models rather than demonstrate features of the language. While new primitives are used, they are not presented in the way they have been previously: you should look them up in the NetLogo dictionary.

9.1 Model 5: Efficiency

The first enhancement highlights the importance of good algorithm design. As built, creating the links takes considerable time. In Snippet 4a, each person separately creates the agentset of potential friends as the code iterates through `ask people [ ]`. Snippet 5 establishes edges with different steps. The new algorithm instead constructs the agentset of potential friends only once, and removes people as their friend spots fill up.

The agentset of potential friends is called 'lonely'. It starts with all people as members. Each person first removes itself from the agentset of potential friends. It then selects the required number of friends from that agentset and creates links with them. If any of the linked people have now met the quota of friends, that person is also removed from 'lonely'.

Model version 5 should initialise noticeably quicker than version 4 (watch how long the setup button stays 'pressed'). Efficiency is particularly important for large models with many agents, where even a small difference in time taken for one agent scales up to considerable time for many agents and time steps. There are tools to identify which procedures are consuming the most computer time either each time they run or simply because they are executed many times (see the profiler extensions at Section 12.3).

As well as the benefits of good algorithm design, this snippet demonstrates the scope of local variables. The agentset named 'lonely' is assigned at the start of the procedure and is then available to all the code at deeper levels in the `ask people` code block until it is deleted automatically at the end of the procedure.

Snippet 5: Faster network creation

1. Rename make-network procedure to make-network-original.
2. Create the new make-network procedure.

```
to make-network
  let lonely people
  ask people
  [ set lonely other lonely
    let needed round num-friends - count my-links
    if needed > 0
    [ let targets n-of min (list needed count lonely) lonely
      create-links-with targets [ hide-link ]
      ask targets [ if count my-links = num-friends
                    [ set lonely other lonely ] ]
    ]
  ]
end
```

9.2 Model 6: Visualising Decisions

Model version 6 enhances the visualisation to display those patches where some specified proportion of people have adopted protective behaviour, providing a barrier to the epidemic spread. This is implemented with an additional breed of turtle agents that provide patch markers. The marker breed is defined and markers created in Snippet 6a. Snippet 6b controls the display of the markers based on the local adoption of protective behaviour. New NetLogo keywords are the primitive `show-turtle` and the built in turtle variables `shape` and `color`.

With these snippets implemented, a small black X will be displayed in those patches where the protected proportion of the people is at least the level set by the slider. As many patches have only 2 or 3 people, a useful level is 0.75 so that a single high attitude person does not trigger the visualisation. You should expect to see a higher density of marked patches near the epidemic frontier, where threat is highest, particularly if the number of friends is set to 0 so that only local threat is perceived.

9.3 Model 7: User-Controlled Perception

Currently, the model constructs an agentset named 'visible' comprising the patch issuing the instruction and its eight neighbouring patches. That patchset is used for the calculations of the social norm (proportion of people protected) and threat. In version 7, the agentset 'visible' is instead a variable owned by patches (implemented in Snippet 7), created during setup to comprise

Snippet 6a: Create markers

1. Define a new breed for the marker called barriers (near top of code).
2. Create the make-barriers procedure (with other procedures).
3. Amend the setup procedure to call the make-barriers procedure.

```
; markers for protected patch
breed [barriers barrier]

to make-barriers
  ask patches
  [ sprout-barriers 1
    [ set shape "x"
      set size 0.7
      set color black
      hide-turtle
    ]
  ]
end
```

the patches within a specified distance. The purpose of storing the visible patchset for each patch is efficiency. The patches in 'visible' never change so the patchset can be created once during setup and stored, which is quicker than the existing approach of creating it each time step (for each patch).

Model version 7 allows the user to control the distance over which the spatial environment influences protective behaviour decisions. Instead of only neighbouring patches, the primitive `in-radius` is used to create patchsets that include all patches within some distance of the patch that owns the variable (see Chapter 11 for detailed discussion of distance). As `in-radius` is a relatively slow operation, the efficiency improvement gained by calculating only once is particularly important.

in-radius

Syntax is **`<agentset>`** `in-radius` **`<given-distance>`**, and selects the agents in the specified agentset (such as patches or turtles) that are within the given distance of the agent issuing the instruction. If the agentset comprises patches, the distance calculation is to the centre of the patch.

9.4 Model 8: Infecting individuals

As represented in this model, the people exist in a world where an epidemic is occurring, but they do not become infected themselves. The epidemic is transmitted through the spatial environment

Snippet 6b: Visualise the markers

1. Create a slider called 'show-protect' (0 to 1 by 0.01, value 0.75) for the proportion of people protected at which the marker will be visible.
2. Create the colour-barriers procedure.
3. Amend the go procedure to call the colour-barriers procedure.

```
to colour-barriers
  ask barriers
  [ if-else count people-here with [ protect? ] /
          count people-here >= show-protect
    [ show-turtle ]
    [ hide-turtle ]
  ]
end
```

rather than from person to person. There are pedagogical benefits of this approach; for example, it facilitates the separate consideration of time, space, and agent characteristics.

This representation was not chosen because of those benefits, however. Idealising transmission as a local environmental process is reasonable for influenza and similar airborne diseases due to the relatively anonymous and incidental interaction sufficient for transmission. This contrasts with blood-borne diseases such as hepatitis or sexually transmitted diseases, where transmission is directly from person to person.

Nevertheless, it may be appropriate to also model the epidemic status of the people agents. For example, you may wish to report the proportions protected for each epidemic state, or to increase perceived threat if a friend becomes infected. This does not require changing the transmission representation. Instead, the local incidence can be used to trigger simulated people becoming infected.

The basic approach is to allow susceptible people on a patch to become infected with probability given by the incidence on that patch (as a proportion of the susceptible population). If all the people have adopted (or all not adopted) protective behaviour, then they should each have the same chance of becoming infected. However, if only some are protected, those people should derive benefit from that behaviour, so the probabilities must be adjusted to average at the appropriate proportion and also recognise the efficacy of protective behaviour. Mathematically, if E is the efficacy of the protective behaviour, P_r the protected proportion of people on the patch, and q the susceptible fraction becoming infected on the patch, the probability of a susceptible person becoming infected is given by:

$$P = \begin{cases} q & P_r = 0, P_r = 1 \\ q\dfrac{1}{1 - P_r E} & \text{unprotected} \\ q\dfrac{1 - E}{1 - P_r E} & \text{protected} \end{cases}$$

Snippet 7: User controlled perception

1. Add a slider called see-distance (0 to 5 by 1, value 2).
2. Add a patch-own variable called visible.
3. Amend the setup-patches procedure to initialise visible (shown).
4. Remove the line that generates the visible patchset from the decide-behaviour procedure.

```
to setup-patches
  ...
  ask patches
  [ ...
    set visible patches in-radius see-distance
  ]
  ...
end
```

As the infection probability depends on the proportion of susceptible becoming infected in that time step, the new procedures to infect people must be called after the spread-epidemic procedure. Once a person has become infected, the probability to recover is the same as used for the patch based epidemic.

Snippet 8 implements infection and recovery, as well as adding monitors to report the proportion infected by protective behaviour status.

The only new primitive introduced is `all?`. This is used to test for the special case of $P_r = 1$, checking whether all people on the patch are protected directly instead of checking whether the proportion protected is 1 (as stated in the equation). The equation for unprotected people will resolve correctly to probability of q when $P_r = 0$ (none protected), so separate checking is not required.

all?

Syntax is `all? <agentset> [<test>]`. The primitive `all?` is a logical operator that reports `true` if every agent in the specified agentset satisfies the test, and `false` otherwise.

This snippet also demonstrates turtles accessing patch variables. The susceptibles agentset has turtle members, so the `ask` shifts NetLogo into the turtle context. Nevertheless, patch variables such as 'incidence' are used directly, there is no need for a construction such as `[incidence] of patch-here`. This capacity for turtles to access the variables owned by the patch where they are located is built into the architecture of NetLogo.

Turtles know about their location

Any turtle is able to access the values of the patch owned variables for the patch where the turtle is located. This access is automatic by simply using the variable name, in exactly the same way as if the turtle owned the variable.

The monitors demonstrate the counterintuitive results that arise in complex system behaviour. A naive approach would expect that the proportion of protected people becoming infected would be lower than the proportion of unprotected people becoming infected, potentially by a factor equivalent to the efficacy of the protective behaviour. With further reflection, however, it is apparent that people near the epidemic front are more likely to adopt protective behaviour (and slower to drop it) and also have greater exposure to the epidemic. Therefore, it is possible that incidence is actually higher in the protected population, making it appear that the protective behaviour increases the risk of becoming infected. Agent-based models can be used to understand such counterintuitive results (see Chapter 2 and Barbrook-Johnson et al., 2017).

The tutorial model is now complete. The final version is displayed at Figure 9.1. Note that the widgets have been organised into related groups. The group headings are Note widgets.

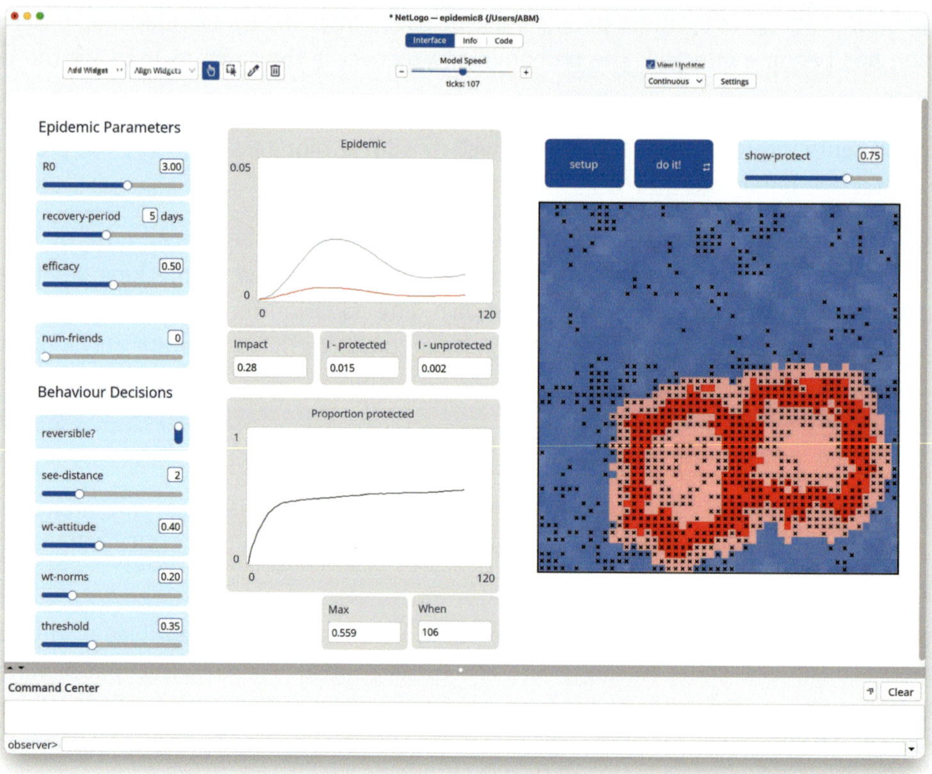

FIGURE 9.1: Screenshot of the interface after completing version 8 of the model. Markers indicate patches where at least 75% of the simulated people are protected.

9.5 Tutorial Progress Check

Model versions 5 to 8 have each introduced an enhancement to the model's features. Very few additional primitives have been used, the features are instead delivered by using previously discussed primitives in new ways. Model 5 substantially decreased the time to construct the network with better algorithm design. Model 6 added a turtle breed to improve the visualisation. Model 7 extended the range of perception through some minor modifications to the patchset construction. Finally, model 8 allowed the people to become infected by accessing the patch based epidemic.

Throughout the tutorial, the instructions have been gradually becoming less detailed. At this point, you should be comfortable with a range of good programming practices such as adding code gradually, checking syntax, interpreting, and resolving some syntax errors, initialising variables, looking up the NetLogo dictionary for useful primitives, saving your model regularly and checking that the code does what you expect. You should also understand what many important NetLogo primitives actually do, and be familiar with common tasks such as adding widgets and letting NetLogo know what variables to create.

There are many other NetLogo primitives that will not be covered. The majority of these are specialised and used very rarely. However, there are two topics that are essential for many models. Any model that includes mobile agents will require primitives that turn, move, look ahead, and implement similar actions (see Chapter 11). Lists have been introduced, but their true power is unexplored. They are required where sequence is important, such as remembering only the most recent interactions (see Section 12.2). There are also specialised sets of primitives that are not part of the core language but are instead added with extensions, such as those to deal with geographic information (see Section 11.6).

9.5.1 NetLogo language

New NetLogo keywords explained in the model versions in this chapter are:

- turtles: `show-turtle`, `shape`, `color`
- agentsets: `in-radius`, `all?`

Snippet 8: Infecting individuals

1. Add a variable for people called state.
2. Initialise the state variable to be "S" when people are created.
3. Create the epidemic-people procedure.
4. Amend the go procedure to call the epidemic-people procedure.
5. Create monitor for the proportion of protected people who are infected. This is calculated using: **count** people **with** [protect? **and** state = **"I"**] / **count** people **with** [protect?]
6. Create monitor for the proportion of unprotected people who are infected.

```
to epidemic-people
  ; infected to recovered
  let infecteds people with [state = "I"]
  if any? infecteds
  [ ask infecteds
    [ if random-float 1 < (1 / recovery-period) [ set state "R" ]
    ]
  ]
  ; susceptible to infected
  let susceptibles people with [state = "S"]
  if any? susceptibles
  [ ask susceptibles
    [ let protect-here count people-here with [protect?] /
                 count people-here
      let infect-here incidence / (incidence + popn-S)
      ifelse all? people-here [protect?]
      [ if random-float 1 < infect-here [ set state "I" ] ]
      [ let prob ifelse-value protect?
        [ infect-here * (1 - efficacy) /
                    (1 - protect-here * efficacy) ]
        [ infect-here / (1 - protect-here * efficacy) ]
        if random-float 1 < prob [ set state "I" ]
      ]
    ]
 ]
end
```

10

Bringing it all Together

The tutorial is now complete. This chapter summarises some of the key ideas presented and points to further materials to improve your understanding of agent-based modelling and NetLogo. It also brings together related ideas that are spread throughout the tutorial. The structure of the tutorial, gradually building a model from start to finish, means that concepts are introduced in the order that they are needed. This chapter, however, is structured thematically, linking together pieces of information that may have been presented far apart.

There are three key skills needed to build an agent-based model. You need to be able to design the model, to understand and describe the entities and behaviour to be represented. You need to use good programming practices to implement the model, so that the model runs quickly and accurately implements the design. Finally, you need to be familiar with the programming language, in this case NetLogo, to translate the intent into instructions. The first two of these skills are required even if you are not personally writing the code; they are necessary to facilitate communication with the programmer.

10.1 Model Design

An agent-based model is a simulation, representing one or more processes with a computer program. That model takes an agent-centric perspective, representing individual entities acting in accordance with their own characteristics and taking into account their own situation.

An agent-centric perspective is critical where there is both heterogeneity and interaction. If the individuals are different in ways that impact on their behaviour, modelling individuals allows that variation to be represented. This contrasts with methods that model the behaviour of an 'average' individual and therefore treat them all as identical and interchangeable. Individuals may interact with each other, choosing their action based on their social environment. Or they may interact with the physical environment, responding to local conditions such as resource availability and altering the environment by their actions. The agent-centric perspective is fundamental to the way the code implements the model design and enables different agents to behave differently in the same situation and the same agent to behave differently in different situations.

This approach is particularly beneficial when modelling complex system behaviour that is driven by the interactions between entities. An agent-based model is able to generate that behaviour from the actions of the simulated individuals rather than attempting to develop rules that directly describe the overall behaviour. That is, the behaviour rules of the agent-based model replicate the theoretical understanding of the behaviour mechanisms that occur in the real world system.

While there is no standard set of questions that will lead to a good model design, there are key features that should be considered. They relate to the representation; identifying the processes, agents, environmental features, and the characteristics and rules that govern interactions. It is important to simplify as much as possible to assist in design and interpretation, but features that are important in the system behaviour must be retained for the model to be useful.

Some potential questions to stimulate a model design are listed at Table 10.1, together with how they apply to the tutorial model. Many other considerations may also be relevant, but these should be sufficient to start any consideration or discussion.

This list of questions identifies key features of a model, but significant effort is needed to create a full model design. In the tutorial, the design and operationalisation was presented as the model was constructed, together with the NetLogo implementation of that operationalisation. In a real project, a relatively detailed design should be specified before any coding. It is not sufficient to specify what is to be included in the model, but also how it is to be included. For example, follow-up questions that focus on the process that is to be represented by the model could include:

- what information is available about how agent attributes and spatial resources are distributed?
- for interactions, what is the specific rule or equation that governs the influence and what information and parameters are required to determine the outcome of that rule?

One important element in model design is distinguishing between the initialisation and the actual simulation. A model starts at a specific point in time, so the initialisation represents a static view of the world (as if time had been paused). Typical tasks in the initialisation include creating the agents and allocating appropriate variable values. These are conceptually different from the simulation of the process as time progresses.

The model design is agnostic with respect to the particular software in which the model is to be implemented. That is, it should be expressed with an agent-centric representation and relevant rules, but not with NetLogo keywords. One way to approach this is to imagine yourself as the agent (each type separately) taking action and think about the choices facing you and the information you require to decide an action. This consideration should be based on theory and data, including narratives and other qualitative information. The goal of the thought experiment is to encourage the agent-centric perspective required for the model specification instead of the perspective of an observer.

Other aspects of the design concern the interface. The input widgets should allow the user to control the primary features of different scenarios related to the research question. Outputs provide the user with information that is relevant to the research question, but also communicate what is occurring within the simulation.

Finally, the design should also consider broader aspects of how the model is to be used. This includes the resolution and accuracy required to answer the research question, how to check whether the model meets the accuracy requirements, and what experiments are to be conducted. These issues are outside the scope of the tutorial, but are covered in other chapters.

TABLE 10.1: Broad design questions for developing the model.

Aspect	Question	Tutorial model
Process	What actions or repeated decisions?	Adopt/drop protective behaviour.
	What state changes without action?	Epidemic transmission. Disease progression.
	What does theory say about these processes?	Behaviour change (Theory of Planned Behaviour, Health Benefits Model, Social Cognitive Theory) and SIR epidemic theory.
Agents	What entities take actions?	People.
	What personal characteristics affect action?	Attitude.
	Any other entities need to be included?	No.
	What characteristics of those entities are relevant?	Not applicable.
	What characteristics change over time automatically?	None.
Spatial	Does location change the way in which the process operates?	Yes.
	What location specific resources (including information) affect action?	Population, viral load.
	How do resources change over time automatically?	SIR equations.
Social	Do other entities change the actions of individuals?	Yes.
	How are agents aware of each other?	Friendship and norms.
AA interaction	What is the interaction between agents?	Pass information about risks, observe norms.
AE interaction	How do the agents affect the environment?	Protective behaviour reduces infections.
EA interaction	How does the spatial environment influence actions?	Infections (viral load) perceived as threat.
EE interaction	How do the characteristics of a location impact on other locations?	Epidemic transmits spatially.

10.2 Good practice in programming

The key good practices in programming agent-based models have been repeatedly emphasised throughout the tutorial; incremental development combined with continual testing. The syntax check only detects problems with the way that the NetLogo code has been constructed; it does not detect errors in translating the model design into code. Such errors can be very difficult to detect in an agent-based model because it is not obvious whether unusual behaviour is due to an error or the complexity of the represented behaviour. The smaller the change, the easier it is to detect these subtle errors.

Each new piece of code should be the smallest change that can be made with which the model will run. It may also be appropriate to program a simpler version of the intended behaviour, and test that, before amending the code to implement more complex behaviour. For example, if you need to have only certain agents take some action, this can be separated into selecting the appropriate agents and then adding the action. Some simpler action (like changing size or colour) could be used to ensure the selection is correct.

The tutorial used many different ways to check whether implemented code worked as expected. Critical to all these methods is to know what to expect in advance, so that your thinking is not affected by the outcome of the test. Test methods demonstrated were:

- Monitors and plots of key variable values and their distribution;
- Colours and other visualisation to highlight relevant behaviour;
- Inspect windows to verify variable values and changes to those values each tick;
- `print` and related statements to verify variable values within a procedure;
- Running specific procedures from the Command Center to check equations are properly implemented.

A real model requires far more extensive testing than demonstrated for the tutorial (see Chapter 14). The tutorial soundness checks are testing only 'face validity'; does the code appear to work as intended?

Another programming practice demonstrated throughout the tutorial is code readability. Factors that contribute to readability include: spacing and indenting, meaningful variable and procedure names, comments (including section headings), and modular code to keep each procedure relatively short.

Modular code is common to many programming languages. Each block of code has a specific task, and can call other blocks of code in a nested structure. In NetLogo, the blocks of code are referred to as procedures, and are delimited with `to` and `end` or with `to-report` and `end`. By convention, the main procedures are given the names *setup* and *go*. These are (usually) triggered by a button, and the code section of the button settings simply calls the relevant main procedure. Each procedure is comprised of calls to other procedures or primitives from the NetLogo language. There is little conceptual difference between a procedure and a primitive; they are simply names for a set of computer instructions, but some are written by the NetLogo developers and built into the language (primitives) while the others are written by the model developer (procedures). Both procedures and primitives are able to use arguments as inputs. Those arguments form the syntax that NetLogo expects when the procedure or primitive is called in the model code.

Finally, good algorithm design is important in making the model efficient, so that it runs as quickly as possible. This becomes critical as the number of agents increases. Using a model will usually involve multiple simulations (see Chapter 13), and any delay in a single run has a substantial impact on the model's long term usefulness.

10.3 NetLogo language

The third thread of the tutorial is NetLogo, which is both a programming language and a modelling environment. The *Code* tab holds the keywords that define the model (such as `breed` and `turtles-own`) and the procedures comprised of commands and reporters that instruct the model's operations. This code tab is automatically linked to the *Interface* tab that supports widgets for model input and output and other model infrastructure such as spatial coordinates.

The tutorial uses only a fraction of the available NetLogo commands and reporters. However, this relatively small set of simple commands can be used to generate rich behaviour. The primitives `with` and `if` are used to select agents that meet specific conditions (see Figure 10.1) and behaviour of the selected agentset is implemented with `ask`. These primitives together allow heterogeneity of characteristics or circumstances to lead to different decisions and behaviour, which is fundamental to agent-based modelling.

Interaction is another essential element of agent-based modelling, so any modelling platform must allow agents to perceive characteristics of other agents. In turn, this requires a concept of 'me' and 'them', as well as the capacity to access the variable values of other agents. In NetLogo, `self` refers to the agent currently in focus, the one to which the instruction is being applied. In contrast, `myself` refers to the agent issuing the instruction. The other key primitive in bridging the agent perception gap is `of`, which is used to specify the information that is known to some agent (or agentset) and is to be perceived by the instructing agent.

Agentsets are comprised of one type of model entity. The World is divided into patches, a regular grid of squares that each cover a unit square of the coordinate system. The entities representing individuals are turtles, and links represent relationships between a pair of turtles (see Figure 10.2). Both turtles and links have `breed` statements to create different varieties. The variables available to each entity are declared in a `patches-own`, `turtles-own`, or `links-own` statement, with different breeds able to declare different sets of variables. These variables are created with the entity to which they belong.

One of the strengths of NetLogo is that any command that makes sense for multiple entity types is identical for all entity types. For example, `ask` iterates through an agentset of patches, turtles, or links, without any need to specify the type. NetLogo also intelligently parses code and determines the entity type to which it applies or expected context.

Entity owned variables are one of three types of variables in NetLogo. Variables are simply names to store and retrieve values. The values that can be stored by variables include text strings, numbers, agentsets, boolean values (`true` and `false`), and lists. Global variables are available

```
patches-own [test?]

to testme
  ask patches [set test? one-of [true false]]
  ask patches
  [ ifelse test?
    [ set pcolor red ]
    [ set pcolor blue ]
  ]
end
```

```
patches-own [test?]

to testme
  ask patches [set test? one-of [true false]]
  ask patches with [test?]  [ set pcolor red ]
  ask patches with [not test?]  [ set pcolor blue ]
end
```

FIGURE 10.1: Each of these code snippets are complete models that set the colour of patches to red or blue based on a randomly allocated patch variable. These demonstrate the two approaches to conditional code with agentsets, testing each agent (**ifelse** within **ask**, upper) and selecting relevant agents in the agentset construction (**with**, lower).

to all model entities (including the observer) and exist for as long as the model is running. These are created either with input widgets on the *Interface* tab or by being named in the `globals` list. Agent variables store information about the specific patch, turtle, or link to which it belongs. There are multiple variables with the same name, each assigned to a particular model entity. An entity is able to access its own variables automatically, and turtles can access the variables belonging to the patch where it is located. Finally, local variables are temporary storage that exist only within the code block in which they are created. They are used, for example, to store the results of a short piece of code to make the model more efficient by avoiding repeating that code or to make the model code more readable to humans. Local variables are created with a `let` statement and the values of all types of variables are updated with a `set` statement.

One of the reasons that NetLogo requires only a small language to implement agent-based models is that many required features are built into the modelling environment, rather than requiring explicit coding. These automatic features include time, space, and the mutual awareness of the code and the interface widgets.

Time is managed by an internal time step counter. That counter is initialised with `reset-ticks` (typically the final command in the *setup* procedure) and advanced with `tick` (typically the final command in the *go* procedure). The tick counter can be observed by any model entity with the `ticks` reporter.

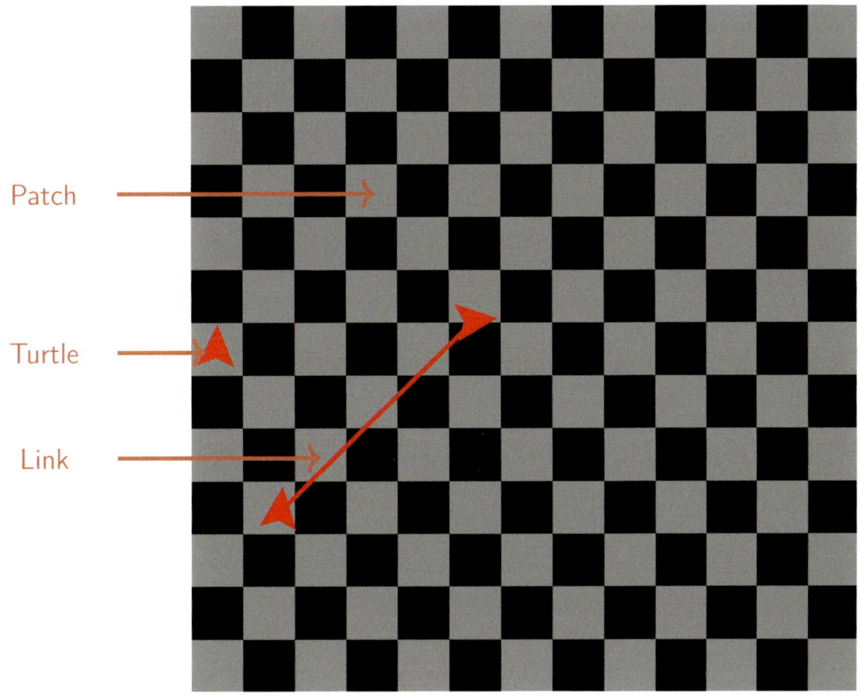

FIGURE 10.2: The three types of model entities available in NetLogo; patches for space, turtles for individuals, and links for relationships between individuals.

Space is dealt with extensively in Chapter 11, as the coordinate system and related concepts such as heading are more salient in models where turtles move. The key spatial concept in the tutorial is `neighbors`, the agentset of patches surrounding the patch or turtle in focus. The `in-radius` primitive extends this spatial neighbourhood further out.

A complete list of NetLogo keywords used in the book appears as a separate index. The following list is the most important keywords that are used in the tutorial, which should become very familiar for any NetLogo modeller:

- procedures: `to`, `to-report`, `end`
- creating entities: `breed`, `sprout`
- variables: `globals`, `patches-own`, `turtles-own`, `let`, `set`
- agentsets: `ask`, `n-of`, `max-n-of`, `with`,
- time and space: `tick`, `ticks`, `neighbors`
- conditional actions: `if`, `ifelse`, `ifelse-value`
- interaction: `myself`, `of`
- mathematical and logical operators

Most of the primitives not covered are for specialised tasks and may never be required. The major exception is those dealing with agent movement (see Chapter 11), which is a common feature of agent-based models. Another important topic is lists (see Section 12.2), which are essential when the order of agents or values matters. For example, if a turtle needs to remember just the most recent five interactions with other turtles, those interactions must be stored in list form so that the oldest can be identified and deleted to make room for the most recent.

10.4 Where to from here?

Congratulations, that's the tutorial all done! The obvious question is what to do next? There is definitely benefit in redoing the tutorial; now you have a framework, you will learn much more deeply.

However, we suggest working through some other materials first to gain a different perspective. At the very least you could work through the tutorial available on the NetLogo website, which will also expose you to some of the movement primitives. In addition, look at some of the models in the *Model Library* and see how they work. Note that these models may use the abbreviated form of some of the primitives. Look up any unfamiliar primitives in the NetLogo dictionary. Many of these library models have suggestions for extensions that you could try to build. Then do the moving turtles tutorial at Chapter 11 before optionally redoing this main tutorial.

The next step is to design and construct your own model. During coding, try and follow the sequencing of the tutorial. Create the agents and their physical world before trying to code the interactions and behaviour. Remember to make the behaviour as simple as possible and then amend it once the simple behaviour works correctly.

At this point, you can start working with some more advanced resources. Many of these are listed at the NetLogo website. You should also dip into the remaining chapters of this book. The most appropriate order will depend on your modelling project and your background.

There are two books that are useful for any agent-based modeller. Despite its name targeting new modellers, *Agent-Based and Individual-Based Modeling: A Practical Introduction* by Railsback and Grimm (2011) is thorough and rigorous and comprehensively covers a range of important topics. While it can be difficult for a complete novice, it uses NetLogo to build the example models and should be accessible after completing the tutorials in this book. A more theoretical perspective is taken in *Agent-Based Models* by Gilbert (2008).

A further two books are particularly good for stimulating ideas about what model to build: *Modeling Behavior in Complex Health Systems* by Keane (2013), and *An Introduction to Agent-Based Modeling: Modeling natural, social and engineered complex systems with NetLogo* by Wilensky and Rand (2015). Both use NetLogo extensively and also have downloadable models available.

Finally, there are developer resources that take a different approach, focusing on the NetLogo language itself and how it is structured. Once you are reasonably comfortable with NetLogo, it is very useful to browse through the *NetLogo Dictionary* and to read the *Programming Guide* thoroughly (included in the online *NetLogo User Manual* linked from the Help menu). These resources will help you become aware of some of the other primitives and gain a deeper understanding of how to think in NetLogo.

Part III

Working with Agent-based Models

11

Turtles on the Move

An important part of many agent-based models is physical space. Often, at least some of the agents move around in their environment, and the physical space influences the actions that the agents can take. For example, livestock roam through grazing areas in land use models, cars or pedestrians must follow routes and interact with each other in traffic models, and random movement can be used to control interactions for economic or social transactions.

This chapter develops a small self contained tutorial model to present the main NetLogo concepts and primitives for mobile turtles. It also uses the GIS code example in the Model Library to introduce more advanced features of including geospatial data (GIS) in NetLogo models to make the physical space more realistic.

11.1 Mobility Tutorial Setup

There are three key concepts to understand for effective use of mobile turtles: the co-ordinate system that identifies location; heading or direction; and moving. The mobility tutorial is designed to demonstrate the basics of each of these concepts. This tutorial progresses much faster than the main tutorial. Only new primitives will be explained, and instructions about creating widgets and creating or amending procedures are brief. There are also only limited reminders of practices such as saving the model and incremental development with testing.

11.1.1 Model Design

The model is a fairly straightforward chasing game. Each turtle will choose another turtle to chase and will chase it until capture. Of course, the turtles being chased are also chasing other turtles. In addition, turtles will be able to detect a chaser that gets too close, and will move away from that chaser as a higher priority than chasing their prey. To make it more interesting, there will be obstacles that stop a turtle from direclty moving toward its target. See Table 11.1 for a translation of this game into responses to the design questions (at Table 10.1).

The general approach to developing this model will follow that of the main tutorial model. Start with creating the world and agents, and then add in the agent behaviour. Consistent with good programming practices of incrementally adding complexity, the chasing behaviour will be implemented (and tested) before any obstacles are included. The instructions are much less detailed than the full tutorial model, and you will need to have completed to at least the end of version 3 to understand how to build the mobility model.

TABLE 11.1: Applying the broad design questions to the chasing game.

Aspect	Question	Tutorial model
Process	What actions or repeated decisions?	Chase and capture.
	What state changes without action?	None.
	What does theory say about these processes?	Move toward target.
Agents	What entities take actions?	Players.
	What personal characteristics affect action?	Speed.
	Any other entities need to be included?	No.
	What characteristics of those entities are relevant?	Not applicable.
	What characteristics change over timc automatically?	Nonc.
Spatial	Does location change the way in which the process operates?	Yes.
	What location specific resources (including information) affect action?	Obstacles to avoid.
	How do resources change over time automatically?	Not applicable.
Social	Do other entities change the actions of individuals?	Yes.
	How are agents aware of each other?	Selected target, avoid any close chaser.
AA interaction	What is the interaction between agents?	Chase target, move away from chaser.
AE interaction	How do the agents affect the environment?	Not applicable.
EA interaction	How does the spatial environment influence actions?	Cannot move through obstacles.
EE interaction	How do the characteristics of a location impact on other locations?	Not applicable.

11.1.2 Setting up the Model

Following the general approach of the main tutorial model, the first steps are to establish the model and initialise the spatial environment. There are several elements that are common to establishing any NetLogo model, setting up the World topology, the *setup* and *go* buttons, and their equivalent procedures. These are contained in Snippet M1, which can be considered a type of template.

Snippet M1: Establish model

1. Open NetLogo (if closed) or create a new model (if open).
2. Amend world settings to 81 by 81 patches (40 in max) and 7 patch size (see Figure 5.1 for hints).
3. Adjust the window size and World location.
4. Create a button labelled 'setup' to call the procedure named 'setup'.
5. Create a button labelled 'step' to call the procedure named 'go' with Forever not checked.
6. Create a button labelled 'go' to call the procedure named 'go' with Forever checked.
7. Enter the following code to create the 'setup' and 'go' procedures.
8. Check the code and Save the model (no further reminders).

```
;--------------------------------------------------
; MAIN CONTROL PROCEDURES
;--------------------------------------------------

to setup
  clear-all
  reset-ticks
end

to go
  tick
end
```

This code includes the standard commands used within the *setup* and *go* procedures. It differs from the starting template used in the tutorial model, where the procedures were created without any content. The tutorial model approach allowed `clear-all` to be introduced with model entities, and `reset-ticks` and `tick` to be introduced with time. However, adding these standard commands at the start of a new model can be a good habit.

The code can be tested by pressing the *setup* button and then either the *step* or *go* button. The tick counter should advance. It is not a complete test, as the `clear-all` could be missing without affecting the model.

The next step is to create the physical environment. However, in the simple version of the behaviour, without obstacles, there is nothing to create as patches automatically have everything necessary. The final initialisation step is to create players, but that first requires an understanding of how NetLogo implements location.

11.2 Co-ordinates

The co-ordinate system within NetLogo was introduced at Section 6.9. All locations are specified with two co-ordinates (*NetLogo 3D* is outside the scope of this book) in the common Cartesian arrangement of an x-axis with values that increase to the right and a y-axis that increases upwards. The intersection of these axes is referred to as the origin. Locations to the left or below the origin have negative values for their x and y co-ordinates, respectively.

Each turtle agent is located at some point in this two-dimensional space. The co-ordinates are stored in the variables `xcor` and `ycor` (x and y axes, respectively), which are automatically created for each turtle. These co-ordinates change as the turtle agent moves (for example, by moving forward some amount). Similarly, setting these variables to specific values moves the turtle agent to the specified location.

Each patch occupies one unit square in the two-dimensional space, centred on integer values of the co-ordinates. The patch boundaries are included in the patch on the left and bottom, and are excluded at the top and right (see Figure 11.1). The patch co-ordinates are stored in the variables `pxcor` and `pycor` (x and y axes, respectively). As patches do not move, these variables can be accessed but not set. Those values are also used to identify patches, so `patch 0 0` is the patch centred on the origin.

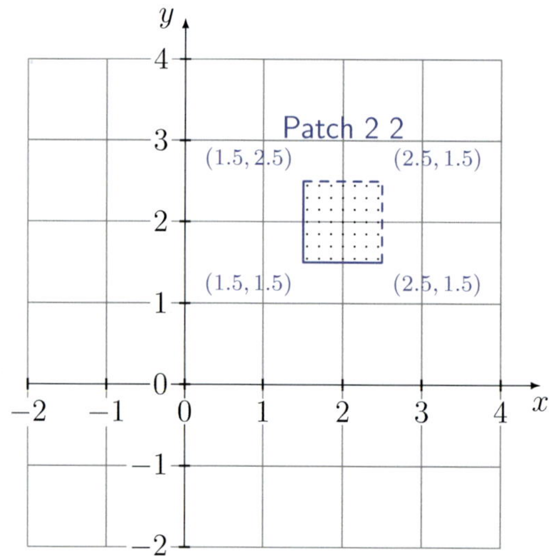

FIGURE 11.1: Overlay of patches with the co-ordinates. The limits of the patch with **pxcor** of 2 and **pycor** of 2 are displayed. Note that the solid lines at the bottom and left indicate that these co-ordinates (**xcor** = 1.5, **ycor** = 1.5 for the patch shown) are within the patch and the dashed line at the top and right (**xcor** = 2.5, **ycor** = 2.5) indicate that these are within the neighbouring patch.

Model settings (see Figure 5.1) control the number of patches in the World and the location of the origin. The only requirement is that `patch 0 0` is somewhere within the World. These settings also control wrapping, whether the edges of the World are borders that cannot be traversed, or whether turtles simply pass through the edge of the World and re-enter at the opposing edge.

11.2.1 Creating players

Having established the environment, the game needs turtle agents to chase each other. From the design, these turtle agents are players that need two variables: speed, and the player they are chasing. That player is to be randomly assigned, the simplest approach. These requirements are implemented in Snippet M2.

At this point, the first new primitives are required. The main tutorial model had each patch `sprout` the turtle agents, creating the appropriate number of people in the centre of each patch. For this model, the observer will create the turtle agents with `create-turtles`.

Players are located throughout the world using `setxy`, which specifies the x and y co-ordinates; with those co-ordinates randomly selected using `random-xcor` and `random-ycor`. Turtle agents have several built-in variables that are created automatically with the turtle. As well as the variables for the co-ordinates (`xcor` and `ycor`), turtles have a `color` and `size`. As for `sprout`, the code contained in the square brackets that is part of the `create-turtles` command is run for each turtle as it is created. That code is used to `set` values for the in-built variables.

Snippet M2: Create players

1. Create a slider for a global variable named num-players, from 0 to 100 by 1.
2. Create a slider for a global variable named max-speed, from 0 to 5 by 0.5.
3. Create a breed of turtles for players with variables named 'speed' and 'target' (place at the top of the code tab).
4. Create the setup-players procedure (place below the setup and go procedures).
5. Add a call to setup-players in the setup procedure (before **reset-ticks**).

```
breed [players player]
players-own
[ speed
  target
]

to setup-players
  create-players num-players
  [ setxy random-xcor random-ycor
    set color blue
    set size 1.5
    set speed random-float max-speed
  ]
  ask players
  [ set target one-of other players
  ]
end
```

setxy

`setxy` `<desired-x-co-ordinate>` `<desired-y-co-ordinate>` places a turtle by specifying the co-ordinates. It is a single primitive version of separately setting the co-ordinates with a combination of `set xcor` `<desired-x-co-ordinate>` and `set ycor` `<desired-y-co-ordinate>`. See also turtle built-in variables `xcor` and `ycor`.

Note that the targets are not assigned until all players have been created. If you tried to assign its target as each player is created, there would be two problems. The first player would have no potential targets, which would generate an error. Also, players created earlier would have more opportunities to be selected as a target and have more chasers than players created later; such a bias is not part of the design.

Press the *setup* button to initialise the game. Open an inspect window for a player and check that it has a target to chase and that the speed is some value up to the maximum speed on the slider.

11.3 Direction

Look at the model after running the *setup* procedure; each player (blue dart shape) is pointing in some direction. This direction is stored in a built-in variable named `heading`, which is the number of degrees clockwise from pointing upwards.

In the same way as other variables, a turtle's heading can be `set` to a specific number, such as 90 to point to the right. But there are other ways to control direction, such as turning `right` or `left` by a specific amount, or facing a model entity or specified co-ordinates. For example, open an inspect window for one of the players and enter `right 90` into the code area at the bottom of the window, the value of the `heading` variable will increase by 90 and the player will turn.

Note that the naming of these commands reflects the agent-centric perspective of NetLogo and agent-based modelling. Turning right or left is unambiguous from your own perspective. However, if somebody else asks you to point right, the question arises as to 'whose right' to point. The reason that headings can be described as clockwise from pointing upwards is that the NetLogo World has a fixed perspective to a model user, a turtle with heading of 0 points upward and there is no way to get behind the computer screen and see the co-ordinate system from the other side.

What happens if `heading` is increased to more than 360 or decreased to less than 0? The value simply adjusts to the equivalent within that range by adding or subtracting 360 as required (that is, `heading` is technically heading modulo 360).

11.3.1 Choosing a heading

Clearly, if the chasing player is to capture their target, the chaser must move toward the target. The most straightforward approach is to face towards the target and then move forward. This is also a natural representation of human behaviour; if you are chasing someone, you would face them (or possibly in front of them) rather than run sideways while facing in a different direction.

However, players are also being chased, potentially by multiple other players. Highest priority is to face away from the closest chaser, and only face the target if there are no chasers in threatening distance. This is implemented with Snippet M3.

Snippet M3: Choose a heading

1. Create a slider for a global variable named 'warning', from 1 to 10 by 1.
2. Add the choose-direction procedure.
3. Amend the go procedure to call choose-direction within an **ask**, as shown.

```
to go
  ask players
  [ choose-direction
  ]
  tick
end

to choose-direction
  set color blue
  let targetme (players in-radius warning) with [target = myself]
  ifelse any? targetme
  [ set color red
    face min-one-of targetme [ distance myself ]
    set heading heading + 180
  ]
  [ face target
  ]
end
```

The player first checks whether there is any nearby player chasing them. If so, they `face` the closest chaser and then add 180 to their `heading`, which faces them in the opposite direction. If there is no close chaser, they `face` their target.

face

The primitive `face` instructs the turtle agent to change its heading so that it is facing the specified model entity, a patch or turtle agent. See also **`facexy`**.

As well as `face`, this code introduces the `distance` primitive, which reports the distance between the asking turtle agent and some other agent (in the official NetLogo sense, so it may be a patch). It is commonly used with `min-one-of` to find the closest turtle agent or patch that meets some requirement. In this case, it finds the closest of all the turtles targeting the asker.

distance

The `distance` primitive reports the distance to the specified turtle agent or (centre of) patch. See also `distancexy` to find the distance to a specific location referenced by x and y co-ordinates.

Finding the closest

The construction `min-one-of` **`<candidates>`** `[ distance myself ]` is used to find the member of the specified agentset that is closest to the asking turtle agent.

Turtle context procedures were shown in Figure 7.1, but were not required in the main tutorial. Here, the *go* procedure has `ask` **`players`** calling choose-direction, which is a turtle procedure. That is, NetLogo is already taking the player's perspective when the code enters the choose-direction procedure. The reason that this approach is used in this model is to ensure that each player makes its full move before moving on to the next player.

Consider what would happen if the *go* procedure called an observer context procedure to choose directions and then an observer context procedure to move (with `ask` **`players`** for each procedure). Every `ask` iterates through the agents. If a player chooses its direction, and then one of the chasers moves within the warning distance before the target has moved, then the target player may move toward a close chaser.

The code can be tested by pressing the *setup* then *step* buttons. Each player will initially choose a random direction, but then rotate to point to its target. You should see this rotation (though it may be difficult against the black background, which will be changed later), and may see a turtle change to red. Inspect a player to identify its target and then check players that it is facing to verify that one of them is the target.

11.4 Movement

The most common primitive used in moving is `forward` (sometimes abbreviated to `fd`), which instructs the turtle to move in whatever direction it is facing. The distance it moves is given in units of the co-ordinate system. For example, if a turtle located at the origin (given by `xcor` of 0 and `ycor` of 0) with a heading of 90 was asked to `forward 1`, it would move to the location with `xcor` of 1 and `ycor` of 0. If, however, the turtle has a diagonal heading, then one unit of distance is not enough to cross the patch.

Other primitives allow a turtle to look at the patch they would move onto if they either moved forward to the next patch or moved some arbitrary heading and distance. Turtles are also able to move in a direction that increases or decreases the value of a specified variable.

11.4.1 Chasing the target

The final step to implement the basic behaviour is for the player to move forward, since it is facing the correct direction. The speed variable controls how far it can move; speed is simply distance per unit time and each tick in a NetLogo model is a unit of time. However, if the chasing player simply moved by the amount available due to its speed, it could overshoot the target. Instead, the player needs to move at maximum speed if the target is too far away, but only as far as the target if it is close. This is implemented in Snippet M4.

Snippet M4: Move toward target

1. Add the move procedure.
2. Amend the go procedure to include a call to the move procedure after the player has chosen their direction (within the player context of the **ask** code block).

```
to move
  let intended-move min (list speed distance target)
  forward intended-move
end
```

forward

The primitive `forward` instructs the turtle agent to move in the direction of its heading by the specified distance. If the World boundaries do not wrap and the requested forward movement would take the agent past the boundary, it stops on an edge patch. It is sometimes seen in code in its abbreviated form, `fd`.

Run the model; you may need to slow it down (speed slider marked (4) at Figure 4.1) to see the players chasing each other. A screenshot is displayed at Figure 11.2.

The chase never ends because nothing happens when a player catches its target. One potential resolution is that the caught player is removed from the game. Any players who were targeting that player then need to choose a new target, but only the player who caught it should be given credit for the capture.

Consider what is required to implement this behaviour before amending the code, as described in Snippet M5. Basic output widgets are also included; you could optionally add plots over time or histograms instead of simply reporting current counts and averages.

The critical line in this snippet is `if distance target < 0.005 [ ]`. If a chaser gets very close to its target then the target is considered to be captured. This is an example of using a tolerance to avoid subtle errors arising from precision limits in the way that computers represent numbers (referred to as floating point errors; see Section 12.4.2).

die

A turtle or link that is asked to `die` removes itself from the model.

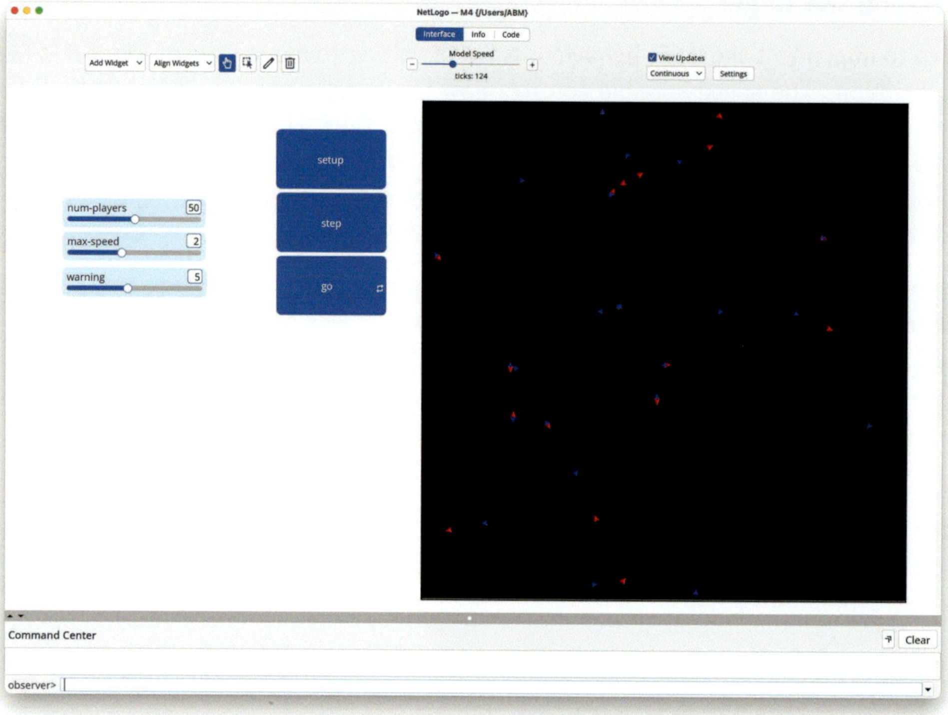

FIGURE 11.2: Screenshot after implementing movement with Snippet M4. The game settles into a pattern of some clusters of players and other players following each other around the World.

Now what happens when you run the model? It still doesn't end (usually), but that's because the last two players are targeting each other. Any time one player gets close enough to almost catch the other, the warning triggers and they run away instead. While a pair of final players is the most common outcome, some simulations will end with three players in a chase, others will have two pairs in separate chases, and sometimes the model will eventually end with an error because one of the two final players is a little faster than the other and catches it and then (unsuccessfully) tries to find another target.

In a real situation, you would need to consider how to determine whether the simulation has finished so it stops. One option is to add a global variable for the number of players still playing. At regular intervals (such as 100 ticks), that variable is checked against the current number of players. If it is the same, then stop the simulation and, if not, replace the variable value. For the tutorial, manual stopping is fine. In addition, adding obstacles may change the outcome because it will be harder for the target to move away.

At this point the model can be considered a full agent-based model, even though it is very simple. Players are individually represented and are 'doing something' through time. Players interact with each other, and those interactions influence the behaviour by chasing and running away. Even without obstacles, the physical environment has an influence because it represents distance. There are several heterogeneous aspects of the players: location, target, and speed.

Snippet M5: Capture target

1. Add a variable for players named 'caught'.
2. Explicitly initialise variable 'caught' to 0 as the player is created.
3. Add the capture procedure.
4. Amend the go procedure to include a call to the capture procedure after the player has moved.
5. Add monitor for number of players.
6. Add monitor for average speed of players.
7. Add monitor for average number caught by players.

```
to capture
  if distance target < 0.005
  [ let my-target target
    ask target [die]
    ask players with [target = my-target]
    [ set target one-of other players
    ]
    set caught caught + 1
    set size size + 0.5
  ]
end
```

11.5 Adding Obstacles

Before adding obstacles, more detailed thought is required about the model design. What is meant by an obstacle? In this model, players will be able to see over an obstacle (that is, they still know how to face their target), but cannot move across it. What is the effect on behaviour? If the player cannot move in its desired direction, then it will instead move in a random direction without an obstacle. Eventually, such random choices combined with the target moving should allow the chaser to move around the obstacle.

The first piece is to create these obstacles in the physical environment. This can be implemented with a boolean variable indicating whether or not a patch is an obstacle, and the obstacles should be visible. The density of the obstacles is to be controlled by the user. These requirements are implemented in Snippet M6.

Now we need to have the players avoid the obstacles. Randomly placing the players runs the risk of having them start on a patch that is actually an obstacle. Snippet M7a avoids this by using a while loop to randomly select a starting location for the newly created player until it is on a patch that is not an obstacle. This snippet introduces the conditional code structure of `while`. The code inside the square brackets runs repeatedly until the condition returns `false`. It also uses the ability of a turtle to directly access the value of variables owned by the patch where it is located (in this case, the value of 'obstacle?').

Snippet M6: Create obstacles

1. Create a slider for a global variable named prop-obstacles, from 0 to 0.2 by 0.05.
2. Create a patch variable named 'obstacle?' (near the top of the code tab).
3. Add the setup-patches procedure (with the other procedures).
4. Add a call to the setup-patches procedure in the setup procedure (before setting up players).

```
patches-own
[ obstacle?
]

to setup-patches
  ask patches
  [ ifelse random-float 1 < prop-obstacles
    [ set obstacle? true
      set pcolor yellow
    ]
    [ set obstacle? false
      set pcolor white
    ]
  ]
end
```

while

The full syntax is `while [<condition>] [<instructions>]`. The instructions are repeated until the condition fails. Note that this repetition is conceptually different from the repetition arising from repeated calls to the go procedure, because the ticks clock does not advance.

You can test this code by pressing the *setup* button and observing the location of players. No player should start on an obstacle (yellow patch). Find a player that appears to be on patch and open an inspect window to check its co-ordinates.

Unfortunately, we cannot use the same trick for avoiding obstacles while chasing because we only know where it is and where it is trying to get to; turtle agents cannot actually 'see' if there's an obstacle somewhere along their intended path. This means there is no obvious way to work out whether the path is blocked.

There are several potentially helpful NetLogo primitives available, of which `patch-ahead` is the most promising. This looks at the patch that is ahead of the asking turtle (that is, in the direction of the turtle's heading) some specified distance. What distance should be checked? The first option is to check the distance intended to be moved, but an obstacle may be in the way without being at the end of the travel. Another option is to check integer points of distance, such as 1, 2, and so on, because each patch occupies one unit square of space. While attractive, this

Snippet M7a: Avoiding obstacles when created

1. Amend the setup-players procedure to add the line with **while**.

```
to setup-players
  create-players num-players
  [ setxy random-xcor random-ycor
    while [obstacle?] [ setxy random-xcor random-ycor ]
    set color blue
    set size 1.5
    set speed random-float max-speed
    set caught 0
  ]
  ask players
  [ set target one-of other players
  ]
end
```

option may not detect a patch whose corner crosses the intended path if that corner is smaller than one unit across. But the idea is sensible; the resolution just needs to be finer than a full spatial unit. The finer the resolution, the smaller the corners that can be detected. This is the approach that is implemented with Snippet M7b.

Sight lines

There is no concept of perception along a line, but NetLogo can be instructed to examine specific points either as absolute positions (x and y co-ordinates) or as a relative position from some anchoring location (heading and distance, or change in x and y co-ordinates). Perceiving along a path must be done by checking multiple points on that path.

If the player detects an obstacle on the path, instead of moving toward its target, it instead randomly selects a neighbouring patch that is not an obstacle and moves one unit of distance onto (or toward) that patch. Limited the distance travelled in this way means that the movement only involves the current patch and the neighbour.

The check-path? procedure implements checking at multiple points. If you can't remember the difference between a command (`to`) and reporter (`to-report`) procedure, review Section 6.8, and look at Section 7.3 for a reminder about the square brackets following the procedure name. The way it checks is that it looks the full distance first and then gradually checks closer. As soon as an obstacle is found, the `while` code block exits and the potential path is rejected.

The model is now complete. The final version is shown at Figure 11.3. You can test the obstacle avoidance by increasing the obstacle density and using the *step* button to move forward one tick

Snippet M7b: Avoiding obstacles when chasing

1. Create the check-path? procedure.
2. Amend the move procedure to call the check-path? procedure and add the alternative movement if the check fails.

```
to move
  let intended-move min (list speed distance target)
  ifelse check-path? intended-move 0.1
  [ forward intended-move ]
  [ face one-of neighbors with [not obstacle?]
    set intended-move min (list 1 intended-move)
    if [not obstacle?] of patch-ahead intended-move
    [ forward intended-move
    ]
  ]
end

to-report check-path? [#how-far #resolution]
  let safe? true
  while [safe? and #how-far > 0]
  [ if [obstacle?] of patch-ahead #how-far
    [ set safe? false
    ]
    set #how-far #how-far - #resolution
  ]
  report safe?
end
```

at a time. Look at a player that is moving toward an obstacle to check that it does not pass through. Alternatively, add a monitor that counts the number of players on patches with obstacles and check that it remains 0.

11.6 Visual Worlds and GIS

Geographic Information Systems (GIS) are the hardware and software to store, manipulate, visualise, and analyse geographic data. GIS data combine location coordinates with information about that location, such as elevation or population density or land use. Locations can be combined to describe the path of geographic features such as rivers and roads. While there are differences, the terms GIS data, geographic data, and geospatial data have sufficient overlap that they are often used interchangeably.

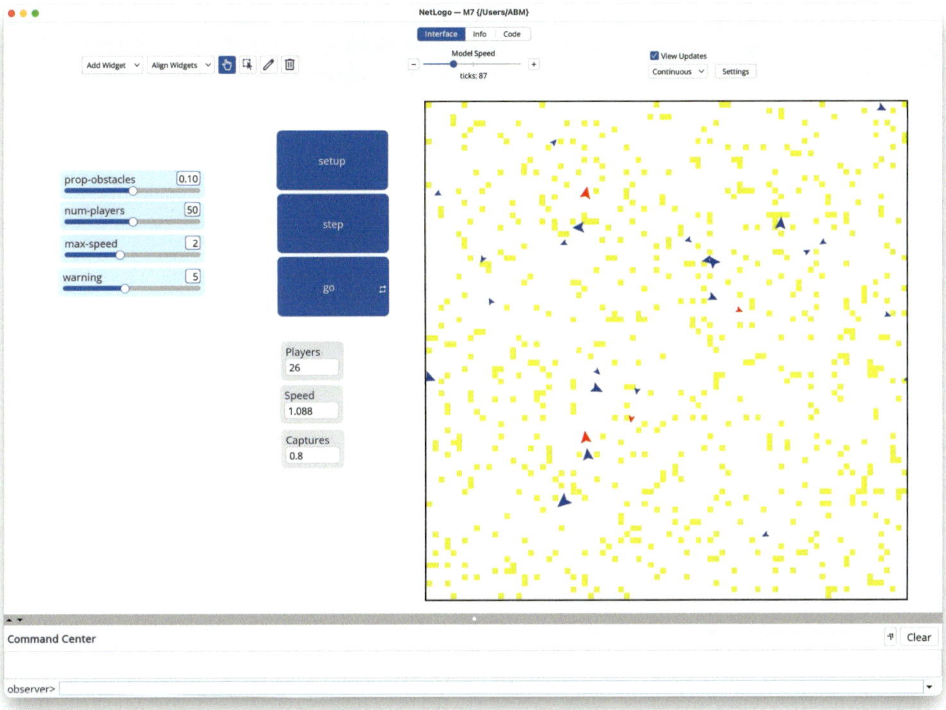

FIGURE 11.3: Screenshot after completing the chasing game model. Players (blue or red darts) avoid obstacles (yellow patches) while chasing each other. Player size indicates the number of captured targets.

NetLogo cannot work directly with GIS data. Instead, GIS data are imported and the information is attached to the standard NetLogo entities of patches and turtles. The commands to do this are included in the GIS extension, which must be loaded at the start of the code.

There are two formats for GIS data. Raster data is laid out as a grid of square cells or pixels (much like NetLogo patches) with a single value for each cell; each specific value applies to the whole of its cell. The dataset also includes the location represented by the corner of the gridded cell data and the real world size represented by the grid. In contrast, vector data is set up like a spreadsheet with a row for each geographic feature that includes both coordinates and attributes for each feature. Layers of vector data can be stacked on top of each other that each provide different information about the same space.

The *gis* folder of the *Extensions Examples* in the Model Library *Code Examples* has two models that demonstrate different aspects of using GIS data within NetLogo. These examples use datasets stored in a subfolder of the NetLogo installation; follow the *models* > *Code Examples* > *Extensions Examples* path using your file manager to access the datasets. The following two simple models each import one of those datasets to demonstrate the differences between raster and vector format.

ncols	360	
nrows	180	
xllcorner	-180	
yllcorner	-90	
cellsize	1	
NODATA_value	-9999	
-3885	-3875	-38
-3408	-3435	-34
-3690	-3690	-36

FIGURE 11.4: Raster meta data

[illegible]	[illegible]	[illegible]	[illegible]
-3810	-3894	-3918	-3
-2212	-2921	-3267	-3
-76	-174	-504	-1
35	27	21	
170	157	177	
809	898	817	
909	944	859	
291	343	314	
440	374	353	

FIGURE 11.5: Raster detail data

11.6.1 Using raster data

Raster data suitable for NetLogo is in ASCII format, with a file extension of *asc*. Two extracts of the example model dataset *world-elevation.asc* are shown, and you can inspect them yourself by opening the dataset in a text editor or spreadsheet software (with space as delimiter). The meta data (Figure 11.4) tells us that the grid size is one degree of longitude by one degree of latitude with 360 columns and 180 rows covering the whole world. It also provides a georeference, the geographic location represented by a corner of the grid. For this dataset, the bottom left corner of the detailed data grid is for 90°S (-90 latitude) and 180°W (-180 longitude), so the value in that cell is the value for the area that runs 90°S to 89°S and 180°W to 179°W. The detailed data (Figure 11.5) displays the values for several of these one degree by one degree sections of the Earth, in this case providing average elevation. Negative values are below sea level.

Snippet GISraster creates a complete model that implements very basic GIS functionality to work with raster data. After building this model, typing 'demo' in the Command Center should produce a recognisable basic map of the world, as shown in Figure 11.6. Note that the GISdata folder must be a subdirectory of the folder where you create the model so that the model can find the raster dataset.

So, what does this code actually do? The first line instructs NetLogo to use the GIS extension. This is a built-in extension; it does not need to be separately installed. However, the GIS primitives within the code will not be recognised without the `extensions` keyword at the top. To use the keywords available from the extension, the call starts with name of the extension and a colon (in this case `gis:`) to let NetLogo know where to find it.

The raster dataset is then read in and stored in the variable named *raster-data*. The connection between NetLogo and GIS occurs in the **gis:apply-raster** command. This takes the stored GIS data and overlays it on the patches, assigning the appropriate value from the raster data to the patch in that position. Remember that the dataset is constructed as 360 columns and 180 rows (coinciding with degrees longitude and latitude), so setting the world size to 181 by 91 means that the overlay almost matches four dataset cells to each NetLogo patch. The **gis:apply-raster** command calculates a value from the raster dataset and assigns that value to the 'elevation' patch variable. These elevation values are then used to colour the patches with the familiar `pcolor` and `ifelse-value` primitives. That is, the GIS information is transferred to the patches and then standard NetLogo code works with that information.

Use `inspect` windows to look at specific patches and see the elevation values. The elevation value for `patch -90 45` is -3885. The four values at the top left of the grid (Figure 11.4) are

Snippet GISraster: Simple GIS raster example

1. Copy the data folder for the GIS extension examples to a more convenient location and rename it to GISdata.
2. Create a new (empty) model and save it at the same location.
3. In settings: 90 for max-pxcor and 45 for max-pycor (with equivalent minimum) and patch size to 4.
4. Add the following code.

```
extensions [ gis ]

patches-own
[ elevation
]

to demo
  clear-all
  let raster-data gis:load-dataset "GISdata/world-elevation.asc"
  gis:apply-raster raster-data elevation
  ask patches [set pcolor ifelse-value elevation > 0 [brown] [blue]]
  reset-ticks
end
```

-3885, -3875, -3408, and -3435. The average of these values is 3650.75. In fact, the value is taken from the closest raster cell rather than averaging over all the raster cells in that patch; that value is -3885, the top left value in the dataset. Different methods to convert the raster values to patch variable values are available but are outside the scope of this book; the important aspect to recognise is that the conversion is not always what you expect and verification is important when working with GIS data.

11.6.2 Using vector data

GIS data is also available in vector format. This is conceptually very different from raster data. Each (geographic) feature in the dataset has both location information (referred to as the geometry) and potentially multiple attributes. In its basic form, the geometry is a point, which might provide the location, name, and population of a city, for example. However, a feature's geometry can have multiple points to mark rivers or roads or other lines. Polygons (series of points with the final point having the same location as the initial first point) are used to describe boundaries such as outlines of countries or patches of land.

The vector format involves multiple files, with the shapefile (extension *shp*) holding the actual data. That data is highly structured but must be converted if it is to be read by a human. Opening the *rivers.shp* dataset with a text editor will show formatted data (Figure 11.7), which can be converted to a readable format such as Figure 11.8 with GIS software (in this case, the *sf* package in the *R* statistical software). The geometry field is a list of pairs of numbers, with each

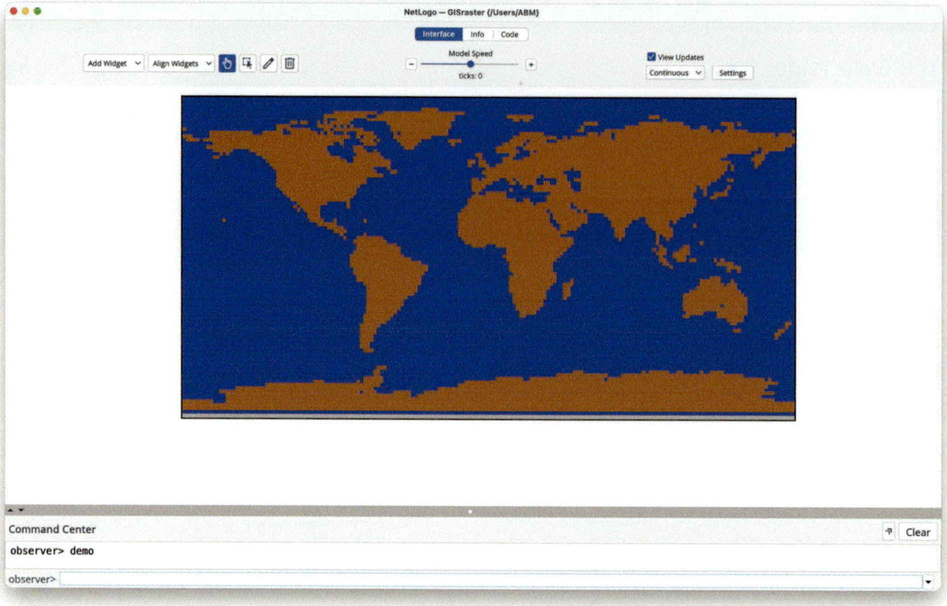

FIGURE 11.6: Map of land and water generated by the simple raster model.

pair providing the coordinates for a point. For this example, the points trace out the route of a river, with separate lists of coordinates for each river in the dataset and some rivers are divided into sections with their own list of coordinates.

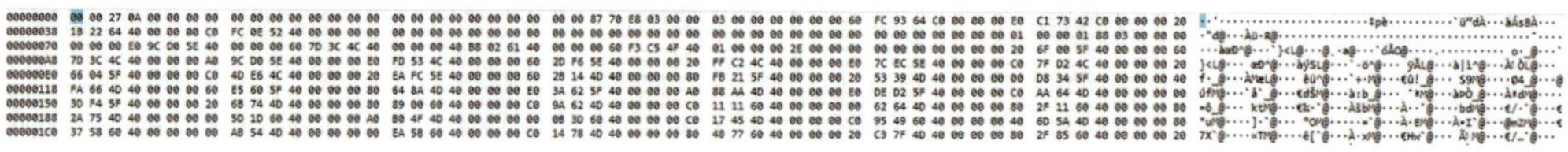

FIGURE 11.7: Raw form of GIS vector shapefile data.

Vector data are more complex than raster data, and the code is less intuitive. Snippet GISvector creates a complete model that imports the rivers dataset and creates a turtle at each point in the GIS line that describes the path of the river.[1] After building this model, typing 'demo' in the Command Center should produce Figure 11.9. Note that the GISdata folder must be a subdirectory of the folder where you create the model so that the model can find the raster dataset.

So, what does this code actually do? As with the raster model, the first line instructs NetLogo to use the GIS extension. The turtles that are to be created at each GIS point need a `breed`, called waypoints in this example. Each waypoint has attributes for the river (name and section) that it belongs to and a sequential number. These waypoints could be used to set out a path for boats to move along, for example, but movement is not included in the example model. Instead, this code simply makes the waypoint turtles visible as white dots.

The vector dataset of rivers is imported and stored in a dataset named vector-data. There are 98 rivers in the dataset (and therefore 98 rows in the interpreted dataset), and the code iterates

[1]The snippet takes advantage of the fact that NetLogo allows multiple commands on the same line to compress the vertical height of this code.

LENGTHCOMP	NAME	SYSTEM	geometry
1822.5530	Aldan	Lena	MULTILINESTRING ((124.0068 ...
2944.2187	Amazon	Amazon	MULTILINESTRING ((-61.27729...
2076.0949	Amu Darya	NA	MULTILINESTRING ((73.98818 ...
3235.0819	Amur	NA	MULTILINESTRING ((122.6396 ...
3170.5035	Angara	NA	MULTILINESTRING ((105.0784 ...
1690.6594	Araguaia	NA	MULTILINESTRING ((-52.84235...
1804.4131	Arkansas	Mississippi	MULTILINESTRING ((-106.1674...
1074.8437	Benue	NA	MULTILINESTRING ((13.73015 ...

FIGURE 11.8: GIS vector shapefile data in human readable form.

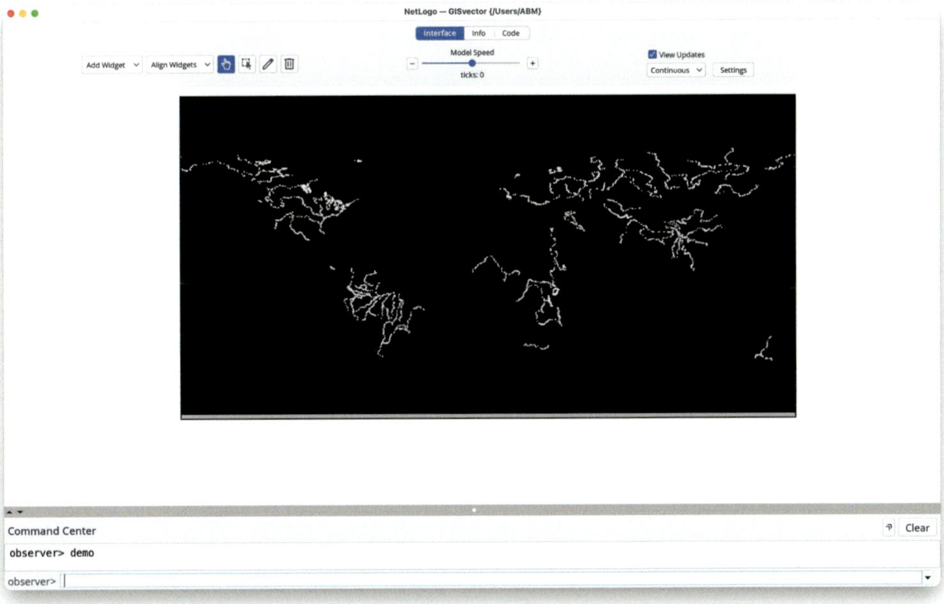

FIGURE 11.9: Map of rivers generated by the simple vector model.

through each river by applying the `foreach` primitive to the list of features (rivers) extracted with the gis:feature-list-of primitive. Each river has one or more sections, each with its own list of vertices (GIS points), so a further iteration is required through those river sections. A third `foreach` iteration steps through each point in the river section and creates a turtle at the relevant location. That location is extracted with gis:location-of.

Lists are discussed in more detail at Section 12.2. The `foreach` command for lists is conceptually similar to the `ask` command for an agentset. Unlike `ask` however, `foreach` iterates through the list members in a specific order, the first item first, then the next item, and so on. The nested `foreach` blocks can make it easy to become confused about which list is being iterated, so good variable names are essential; in this model, the variable holding the list item is named as 'this-' something. For example, when looping though the list of rivers, the current list item is stored in a variable named 'this-river'.

Snippet GISvector: Simple GIS vector example

1. Copy the data folder for the GIS extension examples to a more convenient location and rename it to GISdata.
2. Create a new (empty) model and save it at the same location.
3. In settings: 90 for max-pxcor, 45 for max-pycor, patch size to 4.
4. Add the following code.

```
extensions [ gis ]
breed [ waypoints waypoint ]
waypoints-own [ river   sectionID   sequenceID ]
to demo
  clear-all
  let vector-data gis:load-dataset "GISdata/rivers.shp"
  let idx 1
  foreach gis:feature-list-of vector-data
  [ this-river -> let river-name gis:property-value this-river "NAME"
    let idx-section 1
    foreach gis:vertex-lists-of this-river
    [ this-section -> foreach this-section
      [ this-waypoint -> let this-loc gis:location-of this-waypoint
        create-waypoints 1
        [ setxy item 0 this-loc   item 1 this-loc   set size 1
          set shape "dot"   set color white   set river river-name
          set sectionID idx-section   set sequenceID idx
          set idx idx + 1 ] ]
      set idx-section idx-section + 1 ] ]
  reset-ticks
end
```

foreach

The `foreach` primitive is used to run commands over each item in a list. The full syntax is `foreach` <listname> [temporary-item-name -> code]. That is, `foreach` must be followed by the name of the list from which items are to be extracted and a command block. The command block typically uses an anonymous procedure to pass the specific list member (held in the variable named before the '->' operator, such as the variable named 'this-river') as an argument to the code.

11.6.3 Further GIS explorations

In the GIS vector model, the waypoints are visible as white dots on the map and mark out the rivers. If you combine this with the GIS raster model, those white rivers generated by Snippet GISvector will be in their correct locations on the world map generated by Snippet GISraster. However, this matching of geographic layers is not automatic and occurred here because the two datasets used have the same overall boundaries (the whole surface of the Earth) and projection (the mathematics used to stretch the curved Earth surface onto a flat page). Typically, you will need to set your boundaries (referred to the envelope) and check projections to effectively combine layers of GIS datasets with different types of features.

The two example models focused on creating NetLogo entities from GIS datasets in a way that explained the structure of GIS data. There are other primitives that translate GIS data to NetLogo entities, such as creating turtles directly from points or in the centre of polygons or identifying patches that fall insides polygons. There are also NetLogo primitives to simply draw GIS features in NetLogo worlds. These are described in the NetLogo documentation about the GIS extension, and several are demonstrated in the GIS Code Example in the Model Library.

11.7 Comments

NetLogo has many features built in to support mobile turtle agents. The most fundamental of these is the x and y co-ordinate system, which defines all locations.

A turtle's location is specified by `xcor` and `ycor` variables, its co-ordinate values. A turtle occupies a point only, regardless of the size of the icon used to display the turtle agent. Further, co-ordinates are essentially continuous numbers, so the probability of two turtles being at the same location is negligible unless one turtle is explicitly moved to the other turtle.

One consequence of this arrangement is that NetLogo does not have any built-in method to detect that two turtles are colliding; overlapping icons is not equivalent to occupying the same location. If you wish to check for overlaps, you will need to write your own code to calculate the co-ordinates of the edges of the two icons and check whether the edge of one icon is inside the space occupied by the other icon.

A patch occupies a unit square of the co-ordinate space, spanning from a half smaller (inclusive) to a half larger (exclusive) of the integer co-ordinate value. NetLogo understands the concept of 'here' to mean 'on this patch'.

With a co-ordinate system in place, directions can be defined by reference to that co-ordinate system. Headings in NetLogo are given by a number from 0 to 360 (using modulo arithmetic) with 0 pointing directly up the world and 90 pointing to the right. Headings can be specified explicitly using a number, or indirectly by asking a turtle to `face` a model entity or to change its `heading` by some value.

Distances and movement are also defined by reference to the co-ordinate system. The distance between two points is simply the difference between the co-ordinates. Movement is simply changing location by some distance.

The NetLogo Models Library contains several models that you can examine to help consolidate the concepts and primitives involved in mobile turtle agents. The *Ants* model was briefly discussed in Section 4.2 as an example of agents interacting with their environment. You can now revisit the model to understand how the interaction is implemented. This model is particularly interesting because it uses the `uphill` primitive to instruct ant agents to move in the direction that increases the value of the specified patch variable. As traffic simulation is a common use of mobile agents, you may also wish to examine the *Traffic Grid* model in the library, which uses the co-ordinate system and headings extensively to restrict cars to roads. Note that library models may use the abbreviated form of NetLogo commands, such as `fd` for `forward`.

The library also contains Code Examples to demonstrate coding techniques to achieve particular objectives. This includes an alternative approach to the line of sight problem. Instead of standing at one point and checking at other points along the path, the library model hatches a turtle to move forward gradually, checking the patch where it is located and then dying when it reached the required distance.

This chapter also provided an overview of integrating NetLogo with GIS geographical data. Ultimately, however, geographic modelling is a specialist topic and out of scope of this book. QGIS is open source, multi-platform software for working with GIS data and has excellent documentation (Sutton et al., 2009, including about data formats). There are also several sources of GIS datasets, including NASA's Earth Observation data and official government data sites such as the UK's GeoPortal. There are specific resources about combining GIS and ABM, including some that use NetLogo (Crooks et al., 2018).

Questions

1. Load the NetLogo library model *Rabbits Grass Weeds*. What does the *move* procedure do?
2. Load the NetLogo library model *Traffic 2 Lanes*. Read through the the *move-forward* procedure and look up any unfamiliar primitives in the NetLogo dictionary. How is the code identifying cars that are in the way?
3. Explain the main differences between vector and raster GIS data.

12

Tricks of the Trade

This chapter provides an orientation to some important elements of NetLogo that were not explored sufficiently in the tutorial models. It is a guide to interesting aspects of NetLogo, to introduce you to more advanced material so you understand the potential and what to look for in NetLogo documentation. However, detailed explanation of these topics is beyond the scope of an introductory textbook. We also revisit and extend some good programming practices and discuss how to avoid common errors.

12.1 Code architecture

NetLogo code is modular; it is constructed of nested procedures with each procedure running one or more procedures at the next level down. Each procedure starts with `to` or `to-report` and terminates with `end`. Usually there are two top level procedures, one that calls everything needed to establish and initialise the simulation (conventionally named *setup*) and the other for the procedures that are run during a time step (conventionally named *go*). There is no limit to the number of nested levels; a procedure can call a procedure that calls another procedure that calls another procedure as many levels as necessary to get to basic actions. Where to stop is part of the model developer's judgement. More nesting means the bottom levels can genuinely implement one basic action and nothing else, but the cost is code that is harder to read and debug as related actions become separated.

There are two broad approaches to constructing procedures. The one used in the tutorials in this book takes the observer context. In this structure, the `ask` **`agentset`** is in the same procedure as the actions that the **`agentset`** are to do (see left panel of Figure 12.1). The other approach is to have the `ask` **`agentset [ ]`** in the calling procedure and then the called procedure has the action that is to be undertaken by each agent (see right panel of Figure 12.1).

This is more than just a stylistic difference. The observer orientation (left) is easier to read and debug because `ask` **`agentset`** is in the same procedure as what they are being asked to do, so the whole procedure can be seen together. That is particularly useful for newer modellers because it avoids some subtle errors. However, it has implications for the model logic. There are two separate `ask` statements in the construction on the left, which means all the turtles move then all the turtles eat. In the right-hand construction, a single `ask` statement means that a turtle moves and eats before the next turtle moves and eats. This difference is important if, for example, eating also removes all the food and turtles are only intended to move to where food is available.

There are certain primitives that must be at the top of the code before the procedures start. This includes defining variables with `globals`, `breed`, and **`breed-own`** statements. Below these,

```
to go
  move-turtles
  eat-food
end

to move-turtles ; observer
  ask turtles [ forward 1 ]
end

to eat-food ; observer
  ask turtles
  [ set energy energy + 1 ]
end
```

```
to go
  ask-turtles
  [ move-turtle
    eat-food
  ]
end

to move-turtle ; turtle
  forward 1
end

to eat-food ; turtle
  set energy energy + 1
end
```

FIGURE 12.1: The code on the left has each procedure from observer context, with the **ask** inside the same procedure as the actions the turtles are to take. The code on the right has the actions in separate procedures that are written from the turtle context. This has implications for readability and operation of the code.

the procedures can appear in any order that is convenient; the running order is controlled by the calling procedure (see Figure 12.2). There is no specific practice about such ordering, though most NetLogo programmers would have the *go* and *setup* procedures near the top as they contain the main structure. It is also common practice to put procedures in the same general order as they are called, as that makes it easy to view procedures together that are run sequentially. You may also benefit from including headings (as comments) to separate groups of related procedures.

The Code tab of the NetLogo interface has a drop-down box named 'Procedures' that will jump to the top of any procedure. This can be useful in larger models but relies on the modeller using good procedure names. The name should state what the procedure does. As a simulation is dynamic, procedures typically implement actions, so one way to name procedures is verb then object (for example, 'decide-behaviour').

Laying out the code is a mix between what the computer needs to understand the instructions and what humans need to understand the code (Knuth, 1984). In NetLogo, space is needed around operators and to separate the individual words that make up an instruction but is otherwise irrelevant; that is, you do not need to start a new line for a new command. Multiple commands on the same line is very common for printing information to the screen (such as Figures 8.2 and 12.3) so that all the elements of a line of text to be printed are in the same line of code. Figure 12.3 also has two `let` commands on the same line to demonstrate that such spacing applies more generally, but it also makes the code less readable (see Snippet GISvector for a more extreme examples). Within the constraints of ordering discussed here, code should be as human-readable as possible to reduce errors and aid in verification.

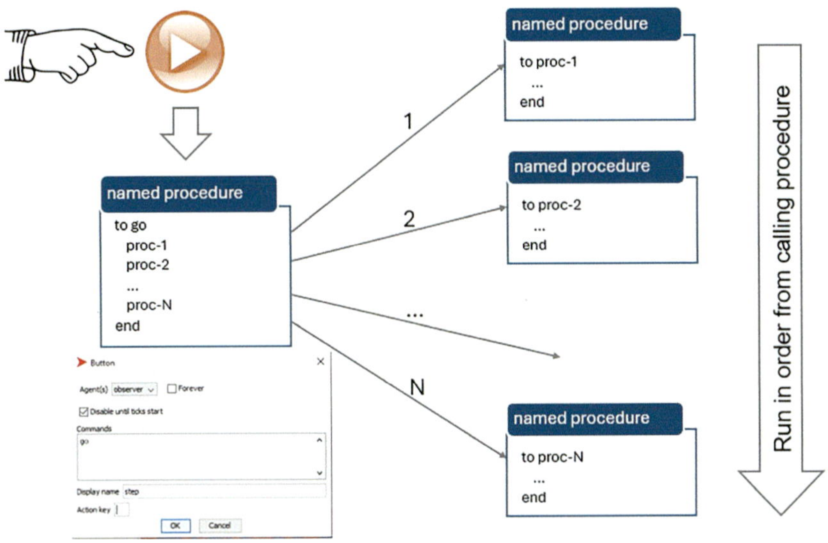

FIGURE 12.2: Each NetLogo procedure implements one element of the process being simulated. These procedures are run in the sequence set out by the procedure that calls them. The highest level procedure is triggered by the model user, typically by pressing a button. That button press runs everything named in any of the procedures in the nested structure.

12.2 Lists in NetLogo

A list is a specific data format; think of it as a container for information; that information may be numbers, strings, agents, other lists, or any mix of these. These are different from agentsets, even if the list contains only agents. An agentset has a single copy of any agent in it (if you add the same agent again, there is no change) and has no order. In contrast, the information in a list has a specific order and can have duplicate entries.

A common use of lists is to store numbers for mathematical operations. Our tutorial model did this at Snippet 1b-3 and Snippet 2e-1, where the maximum (`max`) and total (`sum`) were respectively calculated over the values of a variable from all the agents in an agentset. The reporter `of` constructs a list of values. While that list could have been assigned to a variable and stored (for example with `let mylist [popn] of patches`), both snippets used the list implicitly, making the calculation immediately without storing the list.

For both of these calculations, order does not matter. The maximum or total of a list of numbers is not affected by the order of the numbers in that list.

However, there are situations where order is important. In the tutorial model, new infections were more important in the threat assessment than historical infections, and that was achieved by keeping a running total with a discount. Another way to prioritise recent values would be to average only the most recent values, but that requires NetLogo to identify which values are the recent ones. The need to remember order is an important use of lists.

TABLE 12.1: Conceptual similarities between keywords for agentsets and lists

Action	Agentset	List
Iterate through all members to take some action	ask	foreach
Select a subset that satisfy conditions	with	filter
Select a random subset	n-of	n-of
Retrieve values	of	map
Report number of members	count	length

The code at Figure 12.3 demonstrates manipulation of lists, emphasising the role of fixed order. There are several important aspects to note:

- Two different methods are used for constructing a list: the list named 'weights' simply sets out the values in square brackets, while the list named 'values' uses the explicit `list` keyword. Note that the round brackets are required if the list does not have exactly two members and also that the `list` keyword is inside these brackets.
- Lists in NetLogo are zero indexed. What this means is that the first item is in position 0 (retrieved with `item 0`), which is the same as some other programming languages (like Python or Java) but different from others (like R or Julia). This is seen in the line `set total total + item idx values * item idx weights` since the index named idx is at 0 the first time through the `while` loop, and `item 0` is being read from both lists.
- List keywords use 'first' to mean front of list or item 0 and 'last' to mean end of list, with a left to right reading order. For example, `but-first` means everything except the front item and `lput` means put the new item at the end (last) of the list.
- Lists are immutable. What this means is that you cannot change the contents and must instead create a new list with changes from the old list. For example, the list named 'new-values' is constructed by copying the 'values' list and appending a new member.

This code also uses two keywords that do for lists what different keywords do for agentsets. The primitive `filter` selects only those list members that satisfy some condition, which is similar to `with` for agentsets. The list `length` is its number of members, which is given by `count` for agentsets. Other equivalences are summarised in Table 12.1.

12.2.1 Anonymous procedures

One of the differences between `with` for agentsets and `filter` for lists is that the latter uses an anonymous procedure. Consider the code at Figure 12.4, which is the `filter` from Figure 12.3 reformatted and with comments added. The anonymous procedure is established by the operator `->`, with the name on the left as the variable to temporarily store the current list member, and everything after `->` applied to that list member. This is similar to passing arguments to functions (see Section 7.3).

Anonymous procedures are also used with `foreach` while iterating through lists, as was done when applying `foreach` over a list of GIS data (see Snippet GISvector). This operation is equivalent to `ask` for agentsets; both loop through every member and apply some code to each in turn. For agentsets, the code to be applied is presented in a code block bounded by square brackets as the

```
to testme
  let weights [0.5 0.3 0.1 0.1]
  let values (list 1 2 3 4)
  ; item and spacing
  let total 0 let idx 0
  while [ idx < length values ]
  [ set total total + item idx values * item idx weights
    type idx type "     " print total
    set idx idx + 1
  ]
  type "Weighted average: " print total / sum weights
  ; changing list entries
  type "CurrentValues: " print values
  let new-values lput 5 values
  type "NewValues: " print new-values
  type "NotFirst: " print but-first new-values
  ; list operators
  type "Filtered: "
  print filter [ this-value -> this-value >= 3 ] new-values
  type "ReducedSum: " print reduce + new-values
end
```

```
0     0.5
1     1.1
2     1.4000000000000001
3     1.8000000000000003
Weighted average: 1.8000000000000003
CurrentValues: [1 2 3 4]
NewValues: [1 2 3 4 5]
NotFirst: [2 3 4 5]
Filtered: [3 4 5]
ReducedSum: 15
```

FIGURE 12.3: The code at the top is a complete model that demonstrates some list primitives: **item**, **lput**, **but-first**, **filter**, and **reduce**. The output at the bottom is produced by running this 'testme' procedure.

```
filter                    ; starts the iteration through the list
[ this-value ->           ; current item stored in variable this-value
  this-value >= 3         ; applies the condition and keeps if true
]                         ; ends the iteration
new-values                ; name of the list for the filter
```

FIGURE 12.4: Extract from list manipulation code at Figure 12.3 that uses an anonymous procedure while iterating through a filter.

context changes to the agent. Agentsets also use square brackets to bound the condition when subsetting using `with`.

Anonymous procedures are also used by `reduce`, which is a special list reporter that iterates through the list but implements the code cumulatively rather than separately as is done with `foreach`. That is, the code is applied to the first two items, then to the result and the third item, then to the new result and the fourth item and so on.

The `reduce` reporter in Figure 12.3 (which applies a running total to the list named new-values) uses an anonymous procedure without the operator -> explicitly specifying it. This line is equivalent to `reduce [ [running-total this-value] -> running-total + this-value ] new-values`. Note that two arguments are being passed to the anonymous function; running-total starts at 0 like all variables and keeps track of the result so far, and this-value is the variable storing the current list member as the iteration progresses. The hidden use of anonymous procedures is only suitable for very simple calculations, like the cumulative sum here.

12.2.2 Remembering other agents

Lists are not restricted to storing numbers; they can also contain strings (text), agents (turtles, patches), or a mix of data types. They can even store other lists; for example, the GIS vector data at Snippet GISvector had lists of rivers, with each river stored as a list of river sections, and each river section held a list of GIS points.

If agentsets store agents, why would you ever want to store agents in lists? This gets back to the different structure: lists remember order and also allow duplicates, whereas agentsets do not.

Consider the strategic interaction models described at Section 1.5.2, where each individual in a pair decides whether to cooperate or defect over repeated interactions. In these models, an agent might need to remember what decision their partner made over some time, so they can have complicated strategies like 'cooperate unless there have been at least two defections in the last five decisions'. The agent might also interact with different partners and need to keep track of which of their partners are defecting or cooperating. Then a list of lists could be appropriate, where each member is a list of the identifier for the partner and the action of that partner, perhaps keeping only the last 20 interactions overall. Lists are able to store richer information than agentsets to support more complex strategies.

This does not mean that lists are the only way to achieve these results. There are often many different approaches to implementing any process in NetLogo. If you are storing information about interactions between agents, then it is always useful to consider whether links would be easier. A link can include attributes such as timing (`ticks`) of the most recent interaction, interaction history, or number of interactions.

12.3 Testing for efficiency

If there are many approaches, which one should you choose? There are several factors in this decision. If you are building a model for other people (or your future self) to use and potentially extend, then code readability is very important. While good comments and code architecture are essential, code is also easier to understand if the approach uses a natural representation of the process. On the other hand, different approaches may be more or less computationally efficient. More efficient code is important for those models with a large number of agents or interactions (slowing the model) or where very many experiments are required. We have already extensively discussed representation and readability. But how do you work out whether code is efficient?

In the tutorial model, changing the way in which friends were identified substantially sped up model initialisation (Model 5 at Section 9.1). This was achieved by creating the agentset of potential friends only once and then removing them as they reached their capacity for friendship, rather than re-creating the agentset of potential friends for each agent. For a large enough model, a large enough improvement is noticeable and can be timed manually.

The profiler extension is a more targeted way of improving efficiency. It accurately measures how long it takes for a section of code to run, breaking it down by procedure. This allows you to identify the bottlenecks and focus attention on those procedures where there is the most to gain by redesigning the approach.

The easiest way to understand the profiler is to use it. Make a copy of one of the tutorial models, version 3 or later, and name the copy something appropriate. Add the profiler code to this copy:

- Add `extensions [ profiler ]` at the top of the code.
- Add **profiler:reset** and **profiler:start** in the setup procedure, immediately after the `clear-all` instruction. Note that you want the reset to be before the start so that data from any previous profile runs is cleared.
- Create a button named Profile with two commands: **profiler:stop** and **print** profiler:report as shown in Figure 12.5.

Start and then run the model for a small number of ticks, ending it by pressing the "do it!" button again. The profiler data is collected throughout the run. Press the new Profile button and the report will be printed to the Command Center. The same information is presented three times, in different orders. This information provides the amount of time for each procedure that is run.

There are five columns. The first two are the name of the procedure and how many times it has been called. You would expect to see the procedures used in *setup* only run once and the procedures in *go* to be called multiple times (the same as the number of ticks since the *go* procedure is looped through once per tick). The performance information is in the remaining three columns.

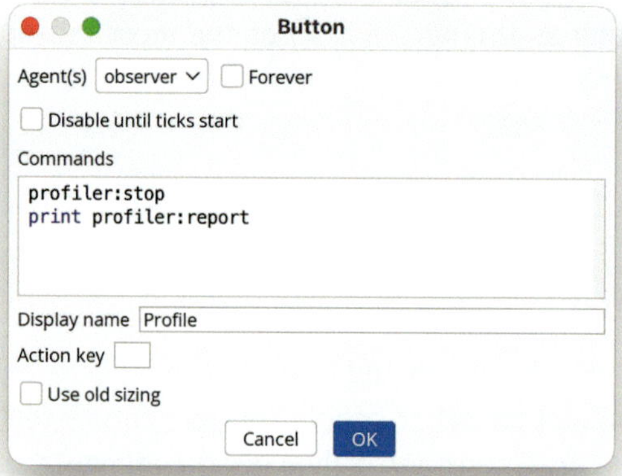

FIGURE 12.5: The button has two commands to end the profiler data collection and then print a report to the console (Command Center).

- Incl T: this is the total time in milliseconds that NetLogo has spent running that procedure and the procedures that it calls. Look at the timing information for the *go* procedure and it will be high because this procedure calls all the other procedures that are implementing the model.
- Excl T: this is the total time in milliseconds that NetLogo has spent running only that procedure, ignoring the time spent in called procedures. This will be small for the *go* procedure as the only time spent in this procedure is the time needed to organise the other procedures.
- Excl/calls: this is Excl T per call to the procedure. That is, it provides the time needed to run through the procedure once (excluding the time spent in called procedures).

The Excl/calls information identifies the most time-consuming procedures in your model. If any of these also needs to run many times, then a different approach to achieving the same outcome could speed up the model substantially. You may also wish to consider whether these expensive procedures can be broken into smaller procedures to help target the specific action that is slowing the simulation.

12.4 Common logic errors (traps for the unwary)

NetLogo has robust syntax error checking (the check button with tick, red backgrounds) that helps to correct errors while constructing code. Online documentation helps with some of the more obscure error messages (Center for Connected Learning & Computer-Based Modeling, 2021), and resolving these is relatively straightforward if you build your model gradually. However, errors in logic are much more difficult to recognise and correct, which is why incremental build and test iteration has been emphasised throughout the tutorial.

```
to setup
  clear-all
  create-turtles 1000 [setxy random-xcor random-ycor]
  reset-ticks
end

to go
  ask turtles [ nextdoor ]
  tick
end

to nextdoor
  ask n-of 500 turtles [ move-to one-of neighbors ]
end
```

FIGURE 12.6: Full model that asks turtles to ask turtles to move. There is a subtle logic error that has each turtle asking 500 turtles to move each tick.

12.4.1 Nested Loops

Isolating a slow procedure is a good way of finding a very common error made by new modellers: nested `ask` statements. NetLogo will run a simulation reasonably well with up to tens of thousands of entities (turtles and patches and links). However, even a small model can easily become impossibly slow by asking all entities to ask all entities to do something. Consider a model with 1000 turtles and you simply want them to move around. If you code it as in Figure 12.6, then 1000 turtles are each asking 500 turtles to move, which is 500 000 movements each tick.

This logic error is very easy to see in this example because there is almost no other code. In a real model, it can be easy to lose track of the level of nesting for `ask` commands. NetLogo offers some protection; if you edit the code to remove `n-of 500` in the *nextdoor* procedure and attempt to run the model, it will terminate with an error. NetLogo ensures that `ask turtles` can only be called in the observer context. This model is in the turtle context when it enters the *nextdoor* procedure as the `ask turtles [ ]` in the *go* procedure changes the context.

12.4.2 Floating-point arithmetic

Another common but subtle error that can be confusing for novice modellers is caused by the way computers store numbers. Numbers are stored as if they have decimal places (referred to as floating point numbers). This can introduce very small inaccuracies in numbers, and therefore errors if you want to test whether two numbers are equal.

You can see this in the output from the code at Figure 12.3: the calculated weighted average is 1.8000000000000003 rather than the correct value of 1.8. Entering this calculation in the Command Centre directly will produce the same result. The correct value is calculated as:

$$0.5 * 1 + 0.3 * 2 + 0.1 * 3 + 0.1 * 4 = 1.8$$

To get the correct answer, remember that precedence rules require the multiplications (*) to be calculated before adding (+) the pieces together (at Section 5.7). One way to fix the precision error is to multiply the weights by 10 and then divide the answer by 10:

$$(5 * 1 + 3 * 2 + 1 * 3 + 1 * 4)/10$$

This produces the correct result of 1.8 because all the numbers used are integers, which can be accurately represented in computers. However, this is not a practical solution, as the numbers used in the model would be confusing to any model user and only work because these numbers happen to have one decimal place.

Instead, try to avoid testing if two numbers are equal. When you are counting things or running other code that will only generate integers then there is no problem. However, if your rules concern more complex calculations with decimal places, then it is safer to allow some small difference in any test. For example, instead of testing whether some total $X = 0.7$, you could test whether `X >= 0.7` for a threshold or whether `abs (X - 0.70) < 0.001` if the equality is important.

12.5 Using the Development Environment

In most programming languages the development environment is separate from the language. That is, you use one piece of software to construct and test the code, but the code is able to run independently of that environment. NetLogo is unusual in that the tools to construct code (Code tab) are integrated with tools to document the model (Info tab) and the interface (Interface tab) that allows the model user to operate the model.

In general, NetLogo models look somewhat clunky; the widgets have a specific look that cannot be changed, and there is a single screen for all the buttons, inputs, world view, and plots. This has some advantages, notably that it helps the user to remember that the model is a simplification.

The limited space for the main model screen also forces the model developer to include only the most important features on that screen. That space should be organised with the input parameters in one area and outputs in another area. Figure 12.7 displays the interface for the *JuSt-Social* model about policy options to manage the COVID-19 epidemic (Badham et al., 2021), demonstrating this style of organisation. There are also headings and descriptions to help the user to use the model effectively.

However, many models require more space than is visible. If you exceed the screen size, scroll bars will appear to allow you or the user to move around an unlimited space. Another feature of the *JuSt-Social* model is visible at the bottom of Figure 12.7: detailed parameters that do not generally need to be adjusted are accessible by scrolling down.

You may also have noticed that the *JuSt-Social* interface is very neat; for example, sliders and boxes line up perfectly and related controls are immediately next to each other with fixed amounts of space between sets of controls. This is difficult to achieve by dragging widgets with the mouse. However, there are tools to automatically align widgets. While developing the model, it is fine to place widgets in reasonable places, but it is worth spending the time to organise the interface properly before presenting it to others.

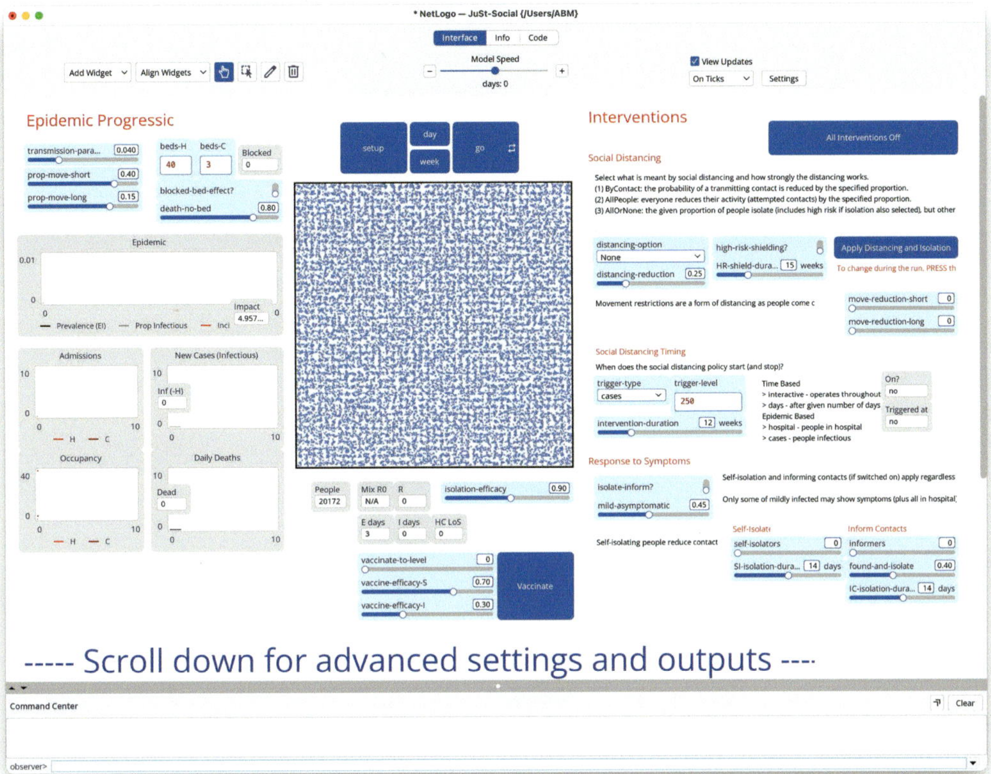

FIGURE 12.7: The JuSt-Social model has areas for epidemic input parameters such as infectivity (top left), epidemic progress reporting (bottom left), buttons to control the simulation (top middle), and descriptions and settings for policy interventions (right).

Questions

1. What is meant by elegant code? Why is it important?
2. What would be reported by the code `show item 1 item 2 (list "X" 4 [3 2 1])`? You can check your answer by entering that code into the Command Center.
3. What can you do if a NetLogo model runs very slowly to find the slowest part of the code?
4. How do you operationalise checking whether a variable value equals 0?
5. What comes automatically with NetLogo as a development environment as compared to just a programming language?

13

Experiments

This chapter is about using your model. You built it to help answer some research question, and the way to answer that question is with simulation experiments and analysis of the simulation output. Experiments are also essential to assess the consistency between the model outputs and empirical data.

Experiments for agent-based modelling take the form of running many simulations, varying the model parameters and observing the effect on the model behaviour. Fortunately, there are tools to help you do that, and much of this chapter is about BehaviorSpace, the batch simulation tool built into NetLogo. Normal data manipulation and visualisation software is used to analyse output, outside the NetLogo environment.

13.1 Measuring Your Model's Output

The first decision is what to measure. Consider the *Segregation* model discussed at Section 1.5.1. Two measures are reported that serve different purposes. This model was developed to demonstrate the hypothesis that a small intolerance can generate substantial segregation. It is easy to observe that clusters of colours consistently emerge in only a small number of ticks simply by running the model a few times. Percent similar is the proportion of neighbours with the same colour, averaged over all agents. This measures the amount of segregation and provides a summary of that observation, which directly relates to the hypothesis. Other measures could also be considered, such as the proportion of agents that do not have any agents of the other colour in their immediate neighbourhood. The other measure, proportion of unhappy agents, is the proportion of agents with too few same colour neighbours compared to the preference level threshold. This helps the user to understand the model.

Playing with the model a few times with different inputs can stimulate ideas for output measures; what patterns or behaviour do you see, and what questions are raised? Having identified the behaviour that must be reported, create a measure that summarises that behaviour. Many of the patterns in an agent-based model are visual, so one way to approach this is to consider how you would describe the pattern to someone who couldn't see the model running. For the segregation model, the dominant feature is the areas of colour, which inspires the proposed measure about agents being surrounded by their own colour.

Tutorial model version 3 implemented agent decisions to adopt or abandon protective behaviour in response to epidemic spread. At this point in the tutorial, friendship links had not been introduced. As such, it is a complete but relatively simple agent-based model and is used as the example for the remainder of this chapter.

The research question inspiring the tutorial model concerns the effect of protective behaviour on the spread of an epidemic. The key measure for this model is therefore *Impact*, the proportion of the agents ever infected. This is already reported by the model. Other potential measures include maximum prevalence, the highest proportion of agents infected at the same time. This is relevant to questions of health system capacity, whether there are sufficient resources to treat all the people needing care. Impact cumulates over time and can be measured at the end of the simulation, at the point it stops because there are no more infectious agents. However, prevalence is zero at the end of the simulation, which means that prevalence needs to be reported at all timesteps (and maximum calculated later), or code could be added to store the maximum to report at the end of the simulation.

It may also be useful to measure protective behaviour. That is needed, for example, to distinguish between two situations in which an epidemic spreads: agents failing to adopt protective behaviour, and spreading despite such adoption. Version 3 of the tutorial model includes a measure of the proportion of agents currently protected. For irreversible behaviour, this measure is sufficient to distinguish between the processes that lead to these two similar outcomes.

13.2 Experimental Design

Running the model a few times and seeing the patterns that emerge may be enough to convince you of some relationship, but how do you demonstrate to other people that the relationship exists? Good experimental design is part of the research process. For the *segregation* model, with enough patience, you could run simulations with many different preference levels and map out a relationship between threshold and measures of segregation.

But the epidemic tutorial model has more parameters, so a more systematic approach is required:

- *wt-attitude* and *wt-norms* are the weights for attitude and social norms respectively in the calculation of desire to be protected. These can be any number between 0 and 1, provided the total is no greater than 1, with the incidence-based threat contributing the remainder.
- *threshold* is the level at which desire to adopt protective behaviour is converted to actual adoption. This is also any number between 0 and 1.
- *reversible?* determines whether that adoption decision can change and the agent ceases the protected behaviour once the desire falls below the adoption threshold.
- *efficacy* concerns the effect of adopting the protective behaviour, the reduction in probability of becoming infected if exposed to the epidemic.
- R_0 and *recovery-period* control how much the epidemic would spread without protective behaviour, and the intensity or speed of that spread.

In theory, if you wanted to understand the effect of these input parameters on the simulation results, you could test all reasonable values for all parameters. That is, run the model with every combination of the possible parameter values and see what happens. However, sweeping the full parameter space is not a practical approach for two reasons.

Parameter Space

Parameter space conceptualises each (input) model parameter as a dimension in abstract space. The potential values of each parameter define the coordinates available for the relevant dimension. Selecting combinations of values for experimentation is also referred to as sampling the parameter space.

The first difficulty is to identify appropriate test values for all parameters. Consider possible values for a continuous variable such as *threshold*. The values must be between 0 and 1, but how many values should be tested? Maybe it is fine to just use 0, 0.25, 0.5, 0.75, and 1, but what if the interesting behaviour happens somewhere that wasn't tested? You can increase the likelihood of finding any interesting behaviour by testing more values, but there is no correct resolution; is it enough to test 0, 0.1, 0.2, and so on all the way to 1, or do you really need to refine further to test 0, 0.01, 0.02, and so on? In contrast, it is straightforward to identify every possibility for categorical (or logical) parameters. For example, there are only two values for *reversible?*, `TRUE` and `FALSE`.

The way to deal with this difficulty is with phased experiments. The first set of values should cover the feasible range (there is no point in testing the value 2 if the maximum possible is 1) but select only a few values. This preliminary stage is like range finding; identify the broad area where the interesting behaviour occurs. Later phases can then target those interesting areas with higher resolution in those selected ranges.

The second difficulty is referred to as the curse of dimensionality. Imagine that you wish to test five values for each of the four parameters that are in the range 0 to 1 (say 0, 0.25, 0.5, 0.75, 1), both logical values for *reversible?*, four values for R_0 (say 1, 2, 3, 5), and three values for *recovery-period* (say 3, 5, 10 days). Testing every set of values involves 15 000 combinations (calculated as $5^4 \times 2 \times 4 \times 3$). In practice, it is not quite that bad for this model, as certain combinations are incompatible; for example, *wt-attitude* of 0.75 and *wt-norms* of 0.75 would have total weight more than 1.

Consider only *threshold* and *efficacy*, two numerical parameters that take values between 0 and 1. Any pair of values for these two parameters can be conceptualised as a point with coordinates given by those values in two-dimensional space. Each parameter adds a dimension to the parameter space and multiplies the number of combinations required by the number of values needed for that parameter. The high number of simulations is due to the number of parameters, seven in this case, and this is a small, simple model.

Each combination of parameters values can be referred to as a scenario, describing the conditions to be tested by simulation. Constructing every combination of the parameter values to be tested is referred to as full factorial experimental design. This is the most comprehensive and thorough way to create scenarios to test, but it is also the most expensive in terms of the number of simulations to run.

Scenario

A scenario is a combination of specific values for every model parameter. That is, each scenario is a point in parameter space.

There are other options that reduce the number of simulations but try to maintain good coverage of the parameter space by efficiently sampling the space. Two common approaches are referred to as 2^k design and Latin hypercube sampling (see Figure 13.1). In 2^k designs, two values are selected for each parameter, one high and one low, and experiments are constructed with every combination of only those values. This results in 2^k combinations, where k is the number of parameters or dimensions. Latin hypercube sampling (McKay et al., 1979) distributes the combinations to be tested throughout the parameter space by selecting each parameter value only once.

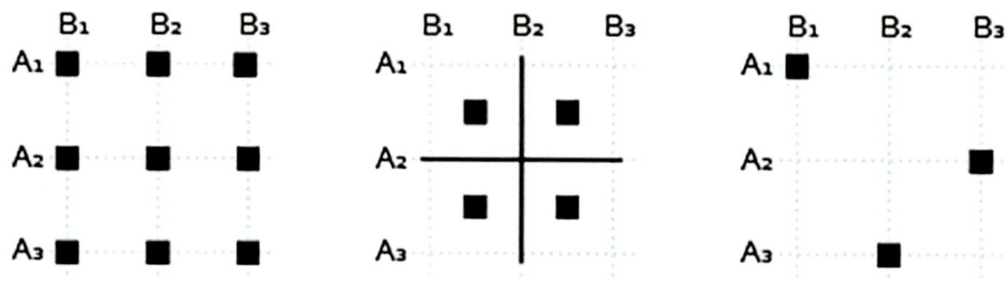

FIGURE 13.1: Three common approaches to selecting experimental parameter values simplified. There are two parameters, with three values each. Full factorial design (left) and 2^k design (central) use every combination, but the latter selects only two values for each parameter. In this example, there are 9 and 4 combinations respectively. Latin hypercube sampling (right) chooses combinations so that each value is used but only once, with only 3 combinations required.

Since the high number of combinations arises more from the number of parameters than the number of values to test for each parameter, a common solution is to drop some variables from the experimental design. How many and which parameters to drop is a matter of judgement, but the key consideration is to retain those that are most relevant to the behaviour of interest. In the tutorial model, that behaviour of interest is adoption of protection behaviour and the consequent effect on epidemic spread.

Behaviour adoption is controlled by the *threshold*; if it is low then agents will adopt protective behaviour with even low levels of threat, social norms, or attitude because the pressure to adopt is likely to be higher than the threshold. If the threshold is very high then only very high levels of threat would be able to trigger adoption, which also means that there is little opportunity for the norms to contribute to adoption. While the two weights (*wt-attitude* and *wt-norms*) contribute to the pressure calculation, varying these weights has a smaller effect on the comparison against a threshold than simply adjusting *threshold*.

R_0 and *recovery-period* both influence the intensity of the epidemic, so it is reasonable to hold one stable and vary only the other. Finally, *efficacy* is a direct effect of protective behaviour on the epidemic, and that behaviour has a stronger influence if it is maintained once adopted.

A reasonable set of parameters for preliminary experiments with the tutorial model therefore varies only *threshold*, R_0 and *efficacy*, holding the other four parameters constant (with *reversible?* set to `FALSE`). Using the same example parameter values, this approach reduces the number of scenarios in full factorial design from 15 000 to only 100 (calculated as $5^2 \times 4$).

13.2.1 How Many Simulations?

Agent-based models are usually random in multiple ways. Initial attribute values and locations of agents are sources of randomness, but there are other hidden sources, such as the order in which agents act. This means that the model must be run multiple times to average out this randomness if you want to make strong claims from the outcomes.

Statistical significance tests are used to consider whether a relationship observed in a sample can be generalised, by estimating the probability that the observed relationship could be an outcome of the randomness rather than 'real'. Ultimately, significance level is related to the amount of variation (or effect size) divided by the square root of the number of simulations. This means that even with only a small amount of variation it is possible to achieve 5% significance (or 1% or whatever target level is desired) simply by running more simulations. Further, any generalisation is about the model; the extent to which the observed relationship in simulation outputs could be generated by randomness in the model rather than arising from change in the parameter values. Generalisation to the system being modelled is about the fidelity of the representation embodied in the model, a question of validation (see Chapter 14). Statistical power is therefore largely meaningless when deciding how many simulations are appropriate.

Rather than approach the question statistically, consider what you want to communicate with the model. If there is a consistent direction in the change of some measure as one parameter value varies, then only a small number of simulations is required for each combination of values because the pattern uses all the simulations. A larger number may be required, however, if the goal is to show that two behaviour choices have different consequences.

One approach is to run a set of simulations with the same parameter values, say 20 runs. Then calculate the variation in the outcome measure of interest from just the first 2 runs, first 3 runs, and so on. The point at which additional runs is not reducing variation is a useful guide for the number of simulations required for each combination. This number may vary for different sets of parameter values, so several test runs may be necessary.

Another relevant factor is the phase of the experiments. When the model is being used for exploration, it is sufficient to have a small number of simulations as the goal is to identify the more interesting parameter ranges. However, more simulations will be required to be confident that the interesting results are valid once focused on the interesting area of parameter space.

For the purposes of demonstrating experimentation with the tutorial model, we use 10 simulations for each parameter combination, which results in 1000 simulations overall. The experimental design is summarised at Table 13.1.

TABLE 13.1: Experimental design for tutorial model.

Parameter	Values	Count
Threshold	0, 0.25, 0.5. 0.75, 1	5
Efficacy	0, 0.25, 0.5. 0.75, 1	5
R_0	1, 2, 3, 5	4
Runs		10
Total runs		1000

13.3 Running Experiments with BehaviorSpace

NetLogo has a built in tool called *BehaviorSpace* to run multiple simulations with specified parameter value combinations. *BehaviorSpace* is easiest to use with full factorial designs as it only requires the possible values to be specified and will implement all combinations. However, it is straightforward to set up multiple experiments if only certain subsets of parameter space are needed and there is also the option to specify subsets of combinations with nested parameter value settings. For more advanced use, it is also possible to run a NetLogo model 'headless' (that is, without the interface) by calling it from a different language such as Java and that allows a different program to control the experiments. Details are beyond the scope of this book but the process is described in the NetLogo documentation.

The example here uses version 3 of the tutorial model. Later versions will take considerably longer to run because they set up many agent links; if you do not have version 3 available, set the num-friends slider to 0 so that these links are not created.

Start *BehaviorSpace* (in the Tools menu) and create a new experiment ('New' button in the dialogue box). A dialogue box will open that is used to describe the set of simulations required. Name the experiment (in the example, it is 'v3experiment') and then look at the box headed 'Vary values as follows (note brackets and quotation marks)'. This is where you specify the experimental design and it is populated with the current values of the sliders and other interface widgets by default. Note that the width and height of the dialogue box are resizeable (by dragging at the edge) but the individual areas within the dialogue box are not.

For the example experiment, first move (cut and paste) the three parameters to be varied in the experimental design (in Table 13.1) to the top of the box. This has no functional effect, but it is good practice to more easily understand what combinations the experiment is running. The input parameter *threshold* is to take the values 0, 0.25, 0.5, 0.75, and 1. This is entered into *BehaviorSpace* as `["threshold" [0 0.25 1]]` as shown in Figure 13.2. That is, the global variable named 'threshold' is to take every value in the interval from 0 to 1 (including the end points), increasing in steps of 0.25. The variable 'efficacy' is varied with the same structure. Note that the sequence structure for the first two variables uses a second set of square brackets. However, the values for 'R0' are simply listed, with a single set of square brackets. The other four variables are also specified, with a single value each as they are fixed for all simulations. The 'Repetitions' box is for the number of simulations to be run for each parameter value combination, in this case 10.

The middle section of the box concerns the measures: what information is the model to report? For the example, we want the impact (proportion of people ever infected) and the proportion of agents who have adopted protective behaviour. Neither of these are recorded as variables in the model so there is no variable to report. However, the *BehaviorSpace* box for reporters also accepts NetLogo code (as shown in Figure 13.3). The NetLogo code (or reporter name) for each measure should be on its own line.

The other decision about *BehaviorSpace* output is how often to generate data. The default is every tick and this would be appropriate if the simulation behaviour during the run is important. However, for this example, only the final outcome is relevant; proportions of people ever infected or

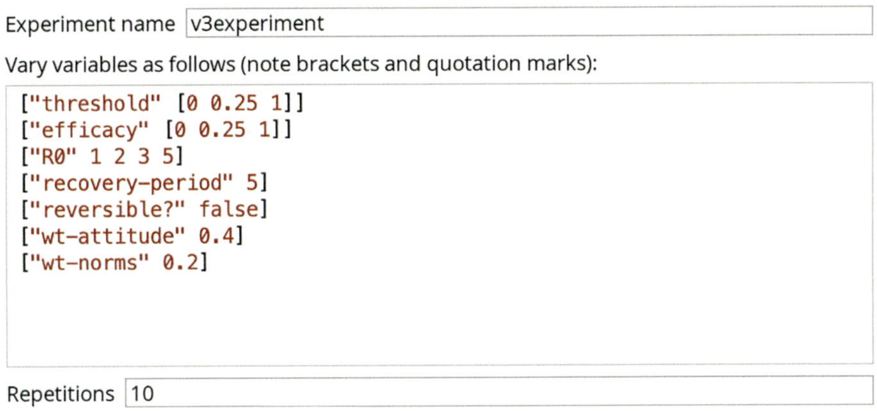

FIGURE 13.2: The top part of the BehaviorSpace experiment description sets out the experimental design. Each variable name is followed by the individual values to be used (for example with 'R0' or 'reversible?') or by defining a sequence of values with the end points and the amount to be varied (for example with 'threshold'). Note the different bracket structure.

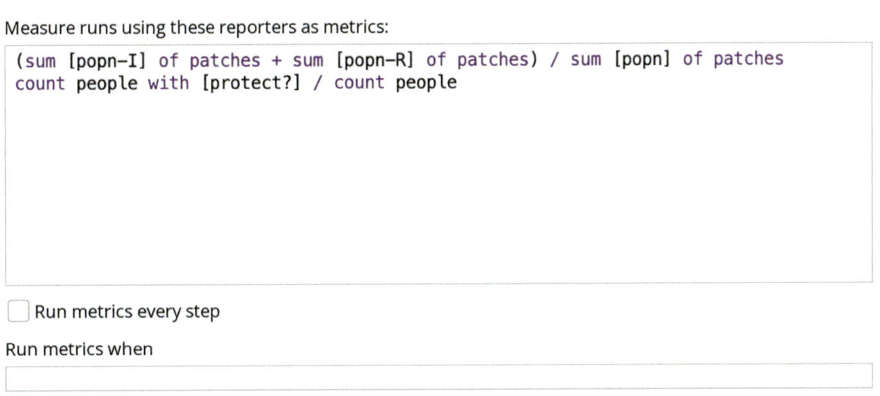

FIGURE 13.3: The middle part of the BehaviorSpace experiment description sets out the measures to be reported and when (every tick, only at the end, or whenever some condition is met).

adopting protective behaviour can only increase and the maximum is achieved when the epidemic ends. Uncheck the box labelled 'Run metrics every step' (see Figure 13.3). There is also a box available to provide NetLogo code to generate output when some condition is met, such as every 10 ticks or if a test fails.

The bottom section of the dialogue box concerns the overall model code (see Figure 13.4). For this example, the initialisation code is contained in a procedure named *setup* and the simulation step code in *go*, following the NetLogo convention. There is also a built-in stop condition, the simulation ends when there are no infectious people to further spread the epidemic. However, all of these aspects can be controlled through specific code in *BehaviorSpace* if necessary. The only control to add for this example is a time limit of 1000 ticks. This is to ensure that every experiment will eventually terminate so that the set of experiments cannot simply run indefinitely.

Once all the experiment description is added, press the 'OK' button and the dialogue box will close. An experiment is now available (Figure 13.5), and the total number of simulations is also

FIGURE 13.4: The bottom part of the BehaviorSpace experiment description identifies the main components of the model and supports contextual code such as when to terminate a simulation.

displayed. As expected, 10 simulations for each combination of parameter values gives 1000 runs overall. Press the 'Run' button and *BehaviorSpace* will ask about the type of format required (Figure 13.6). Table format is generally the most useful, with a suitable format for external analysis. Provide a filename and location for the generated output.

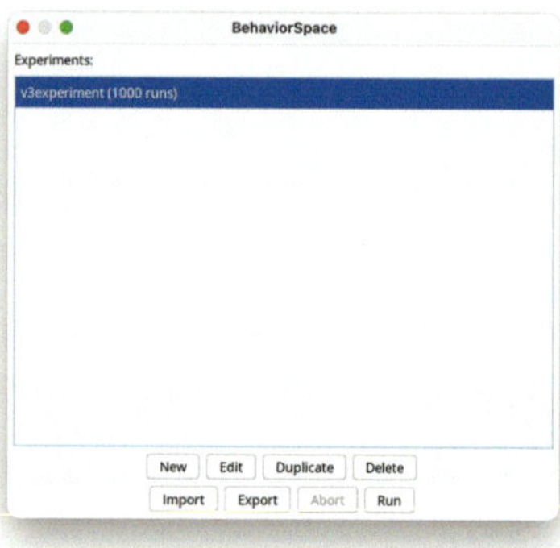

FIGURE 13.5: BehaviorSpace experiment summary with controls to create, edit, and run.

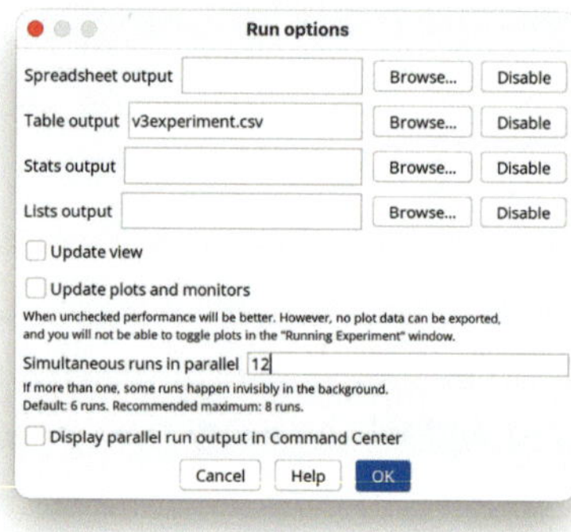

FIGURE 13.6: Dialogue box to control the type of output from BehaviourSpace experiments.

BehaviorSpace automatically takes advantage of multiple cores in the computer, distributing runs to different processors as they become available. You can limit how many cores are being used with the 'Simultaneous runs in parallel' box if you want to reduce the impact on other work on the computer by allocating only some resources to running simulations.

Press the 'OK' button to start the set of experiments. *BehaviorSpace* will report the current run number and tick so you can monitor progress as the experiments are running. It may be useful to occasionally click the 'Update view' or 'Update plots and monitors' checkboxes to see that behaviour is as expected, but displaying model visual output will substantially slow the simulations.

13.4 Analysing Simulation Output

The example *BehaviorSpace* experiments will take some time to run. Eventually, there will be a csv formatted file with the location and filename you specified. Assuming Table output, the format will be:

- 6 lines of metadata including the date and time of the run, and model and experiment name.
- 1000 lines of output, with one line per simulation run.

The simulation output has a specific format with run number in the first column, then the various input parameter values, then the step (usually equivalent to a tick) and then the reported measures. For this example, experiments are only reported at the end of each simulation and there is only one line per run. This means that the 'step' field provides the duration of the epidemic (number of ticks until the epidemic stopped). If measures are reported for multiple steps in each run, the information for each relevant step will appear in a separate row with the same format.

When using multiple processors to run simulations in parallel, all the results will be in the same output file, but the order will be as provided by the processors. The dataset may need to be sorted (by run then step) for visual examination.

While it is important to open and look at the csv file (or spreadsheet equivalent) to check for errors, separate software is usually required for more systematic analysis of *BehaviorSpace* output, such as the statistical software *R* (R Core Team, 2021).

There are three experimental parameters that are varied in the example, and three output values (the two reported measures and duration). Figure 13.7 plots the proportion ever infected (impact) against both *threshold* and *efficacy* in two different ways. Only the plots for $R0$ of 3 are shown, as these were the most interesting from exploratory analysis. The heatmap (left panel) highlights the interaction between *threshold* and *efficacy*, showing that the epidemic is only controlled when *threshold* is low (so people adopt protective behaviour quickly) and *efficacy* is high. Either condition by itself is not sufficient. The line plots (right panel) are easier to observe specific values of impact and also to compare values of *threshold* or *efficacy* but the effect of their interaction is less obvious.

Further simulations and analysis could refine the interaction. A suitable second phase experimental design would focus on the interesting area of parameter space, perhaps holding $R0$ at 3 and varying the other two parameters in the range 0.2 to 0.8 in steps of 0.1 (or even 0.05).

These visualisations tie back to the original research question, identifying conditions for which protective behaviour is able to stifle an epidemic. Other research questions will require different visualisations, particularly those for which the simulation progress is relevant rather than only the end point. Regardless, it is common to investigate the effect of combinations of parameter values because interaction is a core reason to choose agent-based modelling as the research method.

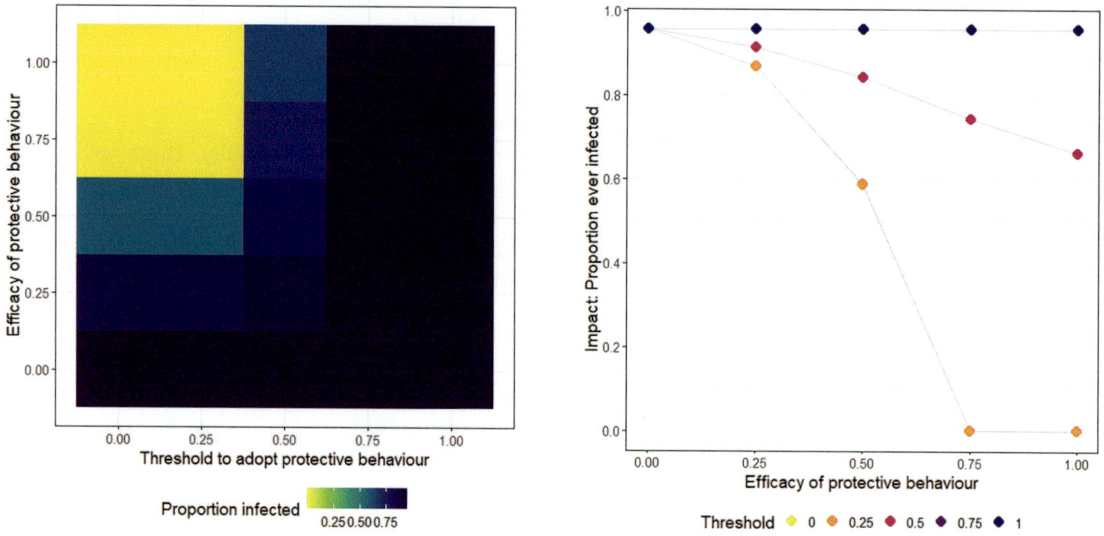

(a) Heatmap: interaction pattern clear. (b) Line plot: value of impact clear.

FIGURE 13.7: Results from BehaviorSpace batch simulation (analysed outside NetLogo) for the experiments with $R0 = 3$. The epidemic is controlled only with both high efficacy protective behaviour and a low threshold for adoption.

13.5 Conclusion

BehaviorSpace makes it easy to create rich sets of experiments. But there is also a risk; patterns may be obscured as you increase the number of varied parameters, making the results difficult to analyse and report. There is no substitute for effective tables and visualisation when trying to communicate relationships and patterns in data. Visualisation is a field of its own and there are many resources available (Wilke, 2019, is an excellent starting point).

The same experimental techniques of batch simulation and analysis are used for several phases of an agent-based modelling project. The chapter framed the experimentation as reporting the answer to the research question, connecting back to the purpose of the model. In more abstract models, the goal may instead be exploration, which also uses systematic experiments so the modeller can examine simulation output for patterns that illuminate the connection between macro-level behaviour and micro-level mechanisms.

For empirical models, another important use of experiments is to calibrate and validate the model (Chapter 14). That is, experiments are used to fit the model to available data as part of the building process and also to assess the quality of that fit. Consequently, experimentation can occur at several points in an modelling project to achieve different goals, and each set of experiments may have different experimental designs.

Questions

Select a NetLogo library model that interests you.

1. What would you need to measure to understand the effect of parameter value changes?
2. Do those measures need to be reported during the simulation, or simply at the end?
3. Which parameters would you vary to comprehensively explore parameter space? What values could you test for each of these parameters?
4. In what ways could you reduce the number of experiments that you would need to run?

14

Is your Model Fit for Purpose?

This chapter looks at the relationship between a model and the real world. There are several aspects to this relationship. Verification checks that the model code is doing what it is supposed to do, appropriately representing your understanding of the processes in the real world. Calibration uses data to set values for the parameters in the model to adequately represent the behaviour that occurs in the real world. Validation assesses whether the model is suitable for the intended purpose. These are the processes that make a model credible, justifying its use for its intended purposes.

Finally, in Section 14.6 we will look at how to describe your model so other people understand (and are persuaded by) your results and how to use the model themselves.

14.1 The Modelling Process (Reprise)

In Section 2.4, we discussed several steps and decisions that need to be made in the process of building an agent-based model. Some of the steps are generic to any kind of research, including defining a research question, and gathering data. There are, however, some aspects that are specific to agent-based modelling, such as questions of abstraction and idealisation, representing processes and the focus on generative causality. These features of agent-based modelling have implications for the relationship with data. For example, a machine learning classifier is assessed based on error rates when classifying data for which the correct class is known but was not used to train it and the process by which the model classifies the data is not relevant, but an agent-based model also needs to show that the process that generates the output data is reasonable. Of course, the model outputs have to resemble the real world, but what resemblance means exactly is harder to define than exact data matching.

It is important to stress that agent-based modelling is an iterative method (Gilbert, 2008). Unless you are building a very simple model, chances are that you will build an initial prototype, test it out, realise there are some important things missing and amend the model. While this might be seen as 'fiddling the books' in quantitative research (meddling with data and introducing bias), it is a normal process for qualitative research (Srivastava and Hopwood, 2009) and theory development (Lave and March, 1993).

For agent-based modelling the iterative nature is intimately connected to our discussion of model boundaries (Section 2.3.2). In trying to understand a complex system the researcher negotiates the model boundaries, trying to ensure that they are not leaving out salient aspects of the system. What is salient, however, can sometimes only be found out by testing a model and realising that it is not (yet!) working.

Given the iterative nature of agent-based modelling, it is important to be completely transparent in reporting the model (see Section 14.6). Verification, calibration, and validation all contribute to this transparency. Importantly, they also operate at multiple levels, which helps to ensure that the model generates outcomes by effectively representing the system rather than coding interesting outcomes into the model. Together, these processes establish the credibility of the model, how much the results should be believed, which is particularly important if the model is to be used to support decision-making (Belfrage et al., 2024).

14.2 Verification

We have already talked briefly about verification (Section 2.4), making sure that the code is sound and internally coherent. That is, verification is the process of making sure that the code does what you intended it to do, that there are no bugs.

One of the best ways to reduce bugs is to write elegant code, so bugs are avoided rather than needing to be corrected later. Elegant code is simple and readable, with useful variable names and comments so it is easy to compare the actual code with the stated intent. Each procedure should only do one task and the task should be described by the procedure name. Sensible formatting also helps with readability, such as positioning the square brackets ([and]) to clearly see the start and end of code blocks. All of these techniques were demonstrated throughout the tutorial model.

Bug

'Bug' is an informal computing term for errors in code. Debugging is the process of correcting those errors.

The tutorial also demonstrated other types of verification, notably the syntax checker and soundness checks. These are most effective when new code is introduced gradually, each new piece implementing the smallest possible change. If you write pages and pages of code without checking, it will be very difficult to debug. In contrast, verifying after each small new addition allows you to focus on the new code as the likely cause of any bug. The syntax check menu button lets you know when the code has straightforward typographical errors (like not closing brackets) that mean NetLogo does not understand the instructions. Soundness checks concern the logic of the code, asking whether it does what was intended. This relies on you understanding what to expect and adding some sort of diagnostic test to confirm that it occurred. For example, the tutorial model had a check for maximum patch population, a histogram to confirm that agent attitudes fitted a triangular distribution, several inspect windows to see the effect of newly introduced code, and `print` commands to report relevant variable values to the command centre.

Some code may pass the syntax checking but then fail when run, leading to NetLogo stopping with an error message (crashing). These 'runtime' errors occur where instructions are correctly formatted but impossible to follow. For example, imagine that you instruct each agent to randomly select an agent on a neighbouring patch and turn it blue (with `ask one-of turtles-on neighbors [set color blue]`), but there are some agents for which there are no agents on neighbouring patches; in this example, the code would be instructing NetLogo to select a random member of an empty set. Another example is where there is a division by zero. Runtime errors might only appear in some runs because they are typically conditional; for example, the empty neighbours issue only occurs when agents are sparse.

Verification

Verification is the process of checking that the model is internally sound.

These conditional bugs can sometimes be found by testing extreme parameter settings of the model. This is referred to as corner case testing because combinations of maximum and minimum fall at the corners of the (multi-dimensional) parameter space. *BehaviorSpace* will help you with that; see Chapter 13.

When debugging, it is essential that you identify the problem and understand what caused the error. Trying out different changes to the code and seeing if it works is akin to guessing; there is no guarantee that you have solved the problem and may be introducing new ones. If possible, isolate the problem by taking even smaller steps. For example, if you added some code to have the agent with the highest value of some attribute perform an action, first check that the agent is being selected properly by colouring it and then check the attribute value is actually higher than for other agents. Once you know the selection works, you can then investigate the action. This approach of identifying and correcting is standard problem solving, not unique to debugging code. The final step is to redo the verification to be sure that the bug has been resolved. It is important to keep in mind that mistakes are common and easy to make, so never be embarrassed by an error in your code.

The interface is another great tool to help verify your model. Some bugs are only evident at the system level. While the code is implemented at the micro- or mechanism level, effects occur at the macro- or system level. After all, the agent-based model is representing a complex system. If something 'looks funny', have a look at what is happening inside the model. Remember the poor entrepreneur that was punished by all of the extorters (Elsenbroich, 2017)? The model ran, it was not throwing up errors, but the processes of the model were clearly not as intended.

Even the most diligent coder might find bugs after the model is supposedly finished. These bugs are often much harder to find and sometimes only show up as strange outputs in the experimentation (as was the case with the over-punished entrepreneur). There might be logic errors in your code, like the wrong referent for a procedure. You might also have used a slightly wrong equation or `ifelse` statement. One major source of bugs is revisions of your model throughout the building process. Every local revision can have effects on other procedures and parts of the code. A code review is a formal process where a different coder checks the model (or sections of it). A fresh pair of eyes can be very helpful.

Verifying an agent-based model has much in common with verification for any other software. All of the methods described are simply good software engineering (Farley, 2022) applied to a simulation project. There are more advanced practices that would be of particular benefit for projects with larger teams. Version control is systematic tracking of who made what code changes when and why; it allows pieces of the code to be added separately by different developers and amended code can be easily reverted if errors are introduced. Unit testing (Thaler and Siebers,

2020) is a more formal way of doing diagnostic testing. Typically, you will verify that new code does what it is supposed to do, but focus on the particular procedure added or changed; unit testing runs the same battery of tests after every change to ensure new code does not have unintended consequences elsewhere in the model.

14.3 Calibration

Even the simplest agent-based model will have a set of parameters; the segregation model discussed in Section 1.5.1 has *happiness*, *density*, and *neighbourhood composition threshold*. Such parameters and their potential ranges are identified during model design and development, referred to as parameterisation. The tutorial model has parameters such as the *recovery-period* for the infection, the *attitude-weight* that defines how much influence the agent's attitude contributes to their behaviour decision, and the *efficacy* of protective behaviour, along with many others. Of these examples, *recovery-period* could take any positive value, but we set it to between 1 and 10 days when creating the slider; *attitude-weight* and *efficacy* are both restricted to the range 0 and 1 because values outside that range are meaningless.

Parameterisation

Parameterisation identifies the parameters for all mechanisms in the model and their feasible ranges.

Having built a model that represents the mechanisms and processes in the system of interest, any model that is not entirely abstract must be calibrated, selecting specific values for all these parameters so that the model is consistent with evidence about how the system behaves in the real world. Before we get to calibration, however, let us briefly look at the different ways real world evidence and data are used in agent-based modelling.

In Section 2.4, we discussed finding out as much as possible about the system we set out to model. All of this information from the real world is used to conceptualise or design the model to effectively represent the target system. Agent-based modelling is particularly good at integrating qualitative and quantitative data (Chattoe-Brown, 2009, 2013). The most common use of qualitative data in designing an agent-based model is for agent behaviour rules. These data can come from observations (describing behaviour options) or interviews (which can add reasons for an agent to execute behaviours and give a deeper insight into decision-making). Quantitative data is mostly used to associate agent attributes with other attributes or with certain decisions or behaviours. In the tutorial we used information from the real world to, for example, initialise cities as areas with high population density (with gradual decrease in density away from the city centre), set reasonable values for the attitude attribute, and include the perception of the threat of illness as an influence in agent decisions to adopt protective behaviours.

There are many quantitative datasets that are useful for agent-based modelling, like census data, national surveys, or cohort studies. They can help us to associate behaviours with particular population groups; for example, younger agents are more likely to be in education and they might also be more likely to go out and meet friends. It can be tempting to use the richness of such datasets to individualise agents, giving them characteristics that match many details in the census data. However, unless these characteristics are *relevant* to the processes being modelled they should be ignored. Following the example, there is no point including age if there is no difference between the behaviours of younger and older agents.

Despite the large amount of existing data, and ever-increasing ways of collecting data (such as sensors and social media streams), you will inevitably come across parts of your model where no data are available. Part of this is about the way data are collected: cross-sectional data from surveys can inform association between characteristics and behaviour but are unlikely to capture mechanisms or the explanation for this association. Even longitudinal surveys typically do not provide mechanisms and reasons, just the amount of change over time and associated characteristics.

There are research designs that elicit situational information and mechanisms, such as simply asking about social context or reasons for change (Barton, 1968). A focus on process is also one of the advantages of small-n research designs for theory development (Smith and Little, 2018). However, it remains true that such data are much less available than the detailed cross-sectional 'status' information that is routinely collected. For example, the developers of an agent-based model of COVID-19 (reported in Chattoe-Brown et al., 2021) found limited data about social contacts; in fact, there are many details in which such data were inadequate for epidemic modelling (Gilbert et al., 2021). One of the recognised uses of agent-based models (and model building generally) is to identify missing information and inform future data collections and research priorities (Barbrook-Johnson et al., 2017).

There may also be data required that is impossible to collect. An agent-based model may have parameters that have no direct correspondence in the real world. These can be seen as model artefacts necessary to make the model function. This was discussed in Section 2.2 in relation to operationalisation. For example, there is no real way to measure the discount factor that was included in the threat calculation for the tutorial model, or even that a discounted count of new cases is the best measure of threat. The weights for each component of the decision calculation (attitude, norms, and threat) are also unmeasurable.

Finally, back to calibration. Calibrating a model sets all the parameters to specific values that best reflect their operation in the real world. If a parameter has a real-world equivalent, the calibrated value can be set directly. For example, the *recovery-period* in the tutorial model could be set to the average recovery period for the specific disease of interest. Where data are not available, either because nobody has measured it or because there is no real-world equivalent to measure, calibration must be done indirectly, either with a proxy or by trying to match the effect on model behaviour and outcomes.

Calibration

Calibration sets values for all parameters of a model to fit model outputs to real-world patterns.

There are two broad approaches to matching model outcomes with empirical data. Traditional optimisation approaches construct one overall measure calculated by the model and use mathematical techniques to systematically adjust parameter values to move that calculated measure toward the real-world equivalent. This has two fundamental problems. The first is that optimising for one measure prioritises that measure over all other aspects of the model and risks overfitting the model to a specific dataset and outcome. While it is possible to construct a combination measure, such a measure requires you to weigh different aspects in advance and hides the trade-offs between the aspects. The second is the problem of equifinality, where different models can lead to the same outputs, so you might get a good fit for the wrong reasons (Williams et al., 2020). Nevertheless, the availability of powerful techniques to focus on the relevant area of the parameter space make this optimisation approach worth considering for some models (McCulloch et al., 2022). We focus on the other main approach, matching multiple patterns.

14.3.1 Pattern Oriented Modelling

Pattern Oriented Modelling is a framework used in agent-based modelling to design, test, and validate models by comparing their outcomes with real-world patterns (Grimm et al., 1996). Importantly, it takes advantage of complexity by shifting the focus from a particular output measure to the interdependence of patterns that arise together from the modelled micro-specifications. Multiple patterns at different scales and of different types are used to constrain and guide model development. The principle is that many pieces of complementary evidence support a stronger claim of effective representation than a close match to a single measure.

The art of pattern oriented modelling lies in identifying suitable patterns. It is important that the patterns are observed for different aspects of the model and are together testing different mechanisms; that is, the patterns should not be correlated. For example, in a model of beech forest development, Rademacher et al. (2004) included behaviour rules for such factors as light, growth rate, mortality rate at different ages, and storm damage. The patterns they used to assess the model included spatial distribution of large trees, similarity of heights within small areas, proportion of trees at different heights across larger areas, and period between maximum canopy cover. This mix of spatial distributions and temporal cycle length (at different scales) is a good way to think about patterns. Other sources of useful patterns are population distributions, response times to some shock, and interactions with the social or physical environment (Gallagher et al., 2021). There are many examples available to help inspire your pattern selection and some guidance on the type of patterns to search for, particularly in the ecological literature (Grimm et al., 2005; Grimm and Railsback, 2012).

What do you do with the patterns once you have found them? For the purposes of calibration, the general approach is to use sets of experiments (see Chapter 13) to explore the parameter space and check which combinations create the patterns of interest. The calibrated parameters are those parameter values used in the experiments where the patterns are best reproduced. Often, patterns are qualitative. In this case, you can use an initial set of scenarios to discard the combinations that fail to reproduce the important patterns of interest and then run a higher resolution set of experiments in the remaining feasible parameter space (Grimm et al., 2005). That is, whether or not a pattern is adequately matched is a yes or no judgement that filters the scenarios to those that are 'good enough' for all patterns. If the pattern fit can be quantitatively assessed or the scenario outputs can at least be ranked by fit for each pattern, then a more systematic approach is available to focus on the more feasible parameter values. In this case, any dominated scenario can be discarded (Badham et al., 2017). A dominated scenario is one where there is a different scenario that is a better fit for every pattern. The retained scenarios are all best in some sense, with each trading off a better fit to one pattern at the cost of a worse fit for a different pattern. The researcher then chooses the output that fits best overall, a judgement call.

Pattern oriented modelling is particularly effective for the indirect calibration needed for those parameters that are embedded in your model that have no clear connection to empirical data. For example, change in many models is driven by thresholds, such as protective behaviour adoption in the tutorial model. These thresholds have to be set in a sensible way but there are no real numbers corresponding to those thresholds in the real world. The further from real-world data a parameter gets, the more important it is that the model is not overly relying on a particular value. For example, if a pandemic model only works when the threat parameter is set to 42 and no other value, something is wrong in the model. This is where robustness and sensitivity play a role; first, however, we move to validation.

14.4 Validation

Validation is the process of assessing whether the agent-based model adequately represents the system of interest; that is, whether it is indeed a model *of* the target system. Models are of different types and have different uses (discussed in Section 2.5), and this has implications for validation. For statistical methods and other mathematical models that summarise a dataset, close numerical matching is essential. The same holds for highly data-driven simulation models, such as microsimulation (see Section 1.4.1). On the other hand, validation for explanatory approaches such as system dynamics (see Section 1.4.3) and agent-based modelling is less about simply replicating numerical data and may emphasise processes and behaviours. Barlas and Carpenter (1990) argue from a more pragmatic and constructivist approach that validity of these kinds of computational models should be linked to usefulness and purpose of the model rather than utopian ideals of objectivity because models are incomplete, functional representations of reality. A model claiming realism and used for policy making has far more stringent accuracy requirements than a model to examine the plausibility of a theoretical mechanism as the cause of an observed social pattern.

You may think that validation has already been completed. After all, verification checks that the code correctly implements the design, and calibration sets parameters to those values that provide the best match between the model and the target system. However, this is not enough to be sure that the representation embodied by the model is of sufficient fidelity to achieve its purpose. The design may be flawed (referred to as structural error); when simplifying the real world, one or more features needed to generate the behaviour of interest may have been lost.

Building with constant iteration between small changes and testing those changes does mean that some validation has occurred while the model is being developed. For example, when you notice some unexpected behaviour, that may be a bug that is easily fixed (an error in coding), but maybe it triggered more detailed thinking about a mechanism and adding an element to the model that you had not initially realised was necessary (an error in design).

Agent-based models represent complex systems, which has consequences for validation. Validation explicitly reconsiders whether the implemented design is a good enough representation of the target system to support extrapolation from model outputs to claims about the real world. It is only when the model is complete that all of the connections exist between the micro-level mechanisms and the macro-system behaviour of interest that is expected to emerge. Some of the relationships already tested may be different after the model is complete, and small changes in inputs may result in large changes in outcomes.

Validation systematically reconsiders the processes that occur within the model and the realism of the outputs. There is no specific procedure that can be used to validate an agent-based model, as the models are used in many different ways (Troost et al., 2023). Instead, validation is more about asking questions about what assumptions have been made, what evidence is available about mechanisms and processes in the real-world system and whether those have been adapted appropriately, and whether the model behaviour is consistent with real-world evidence about how the system responds to changes that can be represented by changes in model inputs (Tesfatsion, 2017; Troost et al., 2023).

Validation

Validation tests whether the model adequately represents the main features, relationships, and processes of the target system.

Many methods for validation of agent-based models are familiar already but applied with a specific perspective. For example, pattern oriented modelling is much broader than a calibration method even though that is where we presented it; it is a way of thinking about design and validation (Grimm et al., 2005). Observing patterns in the system to be modelled guides the features that must be included in the model and the level of detail that may be required to generate those patterns. On the path to finding the best parameter values, calibration explicitly assesses how well the model outputs match observed patterns. If certain patterns cannot be matched, then that suggest the model may not be valid for purposes that rely on a good match. A stronger test holds back some patterns from calibration and assesses the match with outputs from the calibrated model as part of the validation. Methods to support systematic analysis of models include qualitative approaches such as stakeholder assessment and role playing, causal tracing of system behaviours, and data analysis methods such as visualisation and statistical testing (Collins et al., 2024). Selecting an appropriate method will depend on both the type of evidence available and the purpose of the model.

14.5 Robustness: Sensitivity and Uncertainty

There are many sources of uncertainty in an agent-based model. Validation focuses on the fidelity (or lack thereof) of the representation of the real-world process that is being modelled and the potentially limited understanding of that process. Additional to this structural uncertainty are uncertainties inherent to the simulation. These include variation arising from uncertainty about parameter values and randomness between simulations with the same parameter value settings. Experiments (see Chapter 13) are used extensively to understand the robustness of simulation outcomes (Borgonovo et al., 2022).

Uncertainty Analysis

Uncertainty analysis describes the effect of all the uncertainties in the model parameter values by changing the values and observing the effect on the model results.

Using the output from the *BehaviorSpace* experiments of the tutorial model, Figure 14.1 repeats the line plot at Figure 13.7 (right panel) with bars to indicate the highest and lowest values for the 10 relevant simulations (only visible for the scenario with *threshold* of 0.5). These bars highlight the uncertainty arising from random variations in the simulation runs, since the simulations involved have identical scenario settings. Marking extreme values is suitable for a small set of simulations with little variation as for this example, but other exclusion ranges can be also be chosen (such as 5% and 95%, or 25% and 75%).

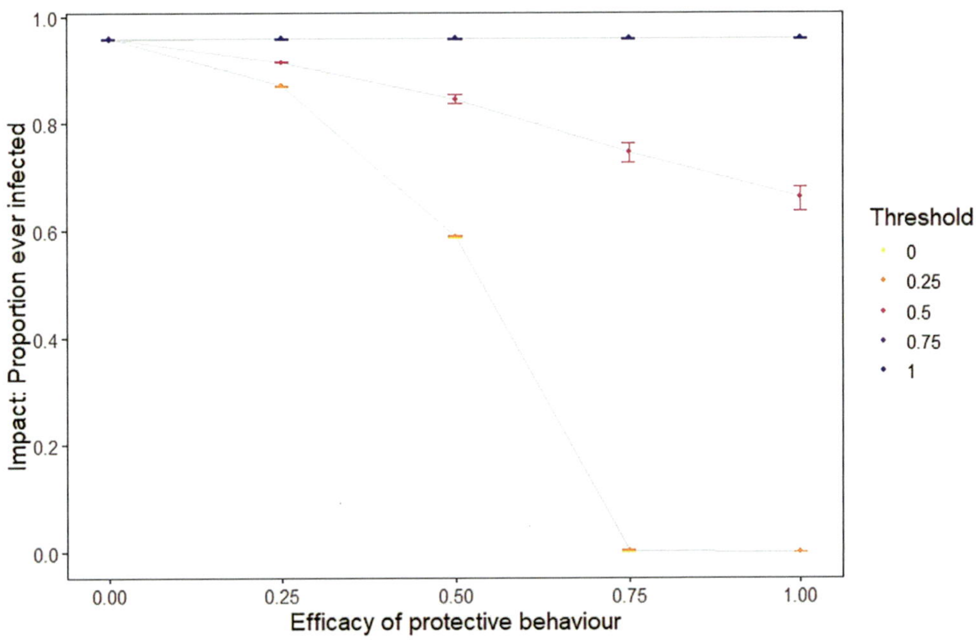

FIGURE 14.1: Results from BehaviorSpace batch simulation (analysed outside NetLogo) for the experiments with $R0 = 3$. Bars indicate uncertainty from randomness, displaying the maximum and minimum measured impact in the simulations for that scenario.

The effect of uncertainty in the parameter values can be directly measured in the simulation output. NetLogo is able to randomly draw from various probability distributions, so it is straightforward to use sliders or other input widgets to specify parameters of the distribution (such as mean and standard deviation for normal distribution) instead of fixed values. Running a large number of experiments will map out the relationship between the probability distributions of the input(s) and the distribution of the measures of interest. This mapping between distributions is what is meant by uncertainty analysis in the wider modelling community, particularly if the model is being used for prediction and decision support.

Sensitivity analysis uses the same techniques as uncertainty analysis, but with a different purpose. For highly empirical models, it is important to understand if the model is sensitive to the particular parameter values chosen. Sensitivity analysis identifies those parameters for which a small error in the calibration substantially changes the model outcome and therefore would support a different decision. Such analysis can be used, for example, to prioritise additional data collection or to design monitoring systems that provide an early warning that other policy options should be reconsidered. These issues are generally not important for abstract models used for exploration and patterns rather than finely tuned projections.

Sensitivity Analysis

Sensitivity analysis measures the importance of the input uncertainties near calibrated parameter values.

Local sensitivity analysis adjusts each parameter by a small amount (10%, for example) one at a time, holding all the others at their calibrated values. This is the easiest approach but misses interaction effects. Global sensitivity takes the opposite extreme and adjusts all parameters simultaneously, which captures interactions but has the same dimensionality difficulty as for experimental design. Worse, for sensitivity analysis, all parameters must be varied rather than just the ones that are more theoretically interesting. The Morris method (Morris, 1991) is a common intermediate approach to explore the parameter space with interactions but with a smaller number of simulations. A small number of random paths are constructed through the parameter space, each starting from a scenario near the calibrated baseline scenario and then changing one parameter, then another, and so on in a random order.

14.6 Application and Communication

Having extracted the important lessons from the model, the next step is to publish the research. Your model is only fit for purpose if you manage to communicate it and use it for the intended purpose. As always, how to communicate an agent-based model depends on the kind of model you have built. An abstract theoretical model, like the Bounded Confidence model (Hegselmann and Krause, 2002) will need the processes to be described in detail but will need little empirical grounding, whereas a model about land use or the housing market needs to have good overlap with the real world and be firmly embedded in existing research.

14.6.1 Protocols for Reporting Agent-based Models

There is no fixed format for the publication of agent-based modelling, despite attempts to standardise the process. One well established framework is the ODD (Overview, Design concepts, and Details) protocol (Grimm et al., 2006, 2010, 2020), which sets out a structure and specific content for describing agent-based models.

The benefits of the ODD are that it improves clarity and standardisation in agent-based model descriptions and aids with transparency of the model. This makes it easier to evaluate the models, facilitates comparison between models, and enhances reproducibility. For highly empirical models, ODD is an excellent protocol. It is widely used in ecology modelling and other applied disciplines (Anshuka et al., 2022). However, it is less widely used in other disciplines.

TABLE 14.1: The Overview, Design Concepts, and Details (ODD) Protocol

Level	Aspect	Action
Overview	Purpose and patterns	Discuss why the model was created.
	Entities, State Variables, and Scales	Define agents, their attributes, and time/space scales.
	Process Overview and Scheduling	Describe model dynamics and execution order.
Design concepts	Design Concepts	Explain the theoretical foundations and mechanisms driving agent behaviour. Includes concepts like emergence, adaptation, fitness, and interactions.
Details	Initialisation	How the model starts (e.g., initial agent distribution).
	Input Data	External data sources, if any.
	Submodels	Descriptions of internal model components.

Tesfatsion (2017) advances three criticisms of the ODD protocol. First, the extensive reporting requirements of the protocol exceed the normal page limits for many journals. Second, the protocol separates agent attributes from data and methods (behaviours), but considering these together is a more natural understanding of 'agent'. This has implications for some programming languages, but more relevant here is that this separation can contribute to unnatural descriptions of mechanisms. Finally, the details asked in the three last reporting points of the ODD are not suitable for general readers. That is, ODD is about reproducibility, describing a model so that other modellers can understand, replicate, and adapt it.

Other types of audience may have different requirements. For example, documentation for a policy model would need to include 'user instructions' such as what the widgets do and how to run the model. Such documentation may also include example scenarios that gradually add more model features, to establish credibility with simple scenarios and help the user understand how to set up the scenarios of interest to them. ODD can also be cumbersome for simple theoretical models, as too many irrelevant details are required.

The needs of different audiences is also addressed by Müller et al. (2014), which starts from the premise that model descriptions serve multiple overlapping purposes and suggests that no single ideal standard exists because different contexts demand different details. Different descriptions are essential for communicating the model to diverse audiences and for diverse modelling purposes. The authors suggest nonetheless that striving for increased standardisation would help transparency and improve information exchange, support theory building, and enhance the capacity of agent-based models to address complex human–natural system challenges.

The original authors of the ODD have continued developing the protocol, including a version focused on visualisation (Szangolies et al., 2024). The vODD provides templates which help to communicate the ingredients of the model, including entities, environment, and processes.

Dilaver and Gilbert (2023) take a slightly different approach to the question of how to communicate agent-based models. Rather than starting from a complexity science perspective, they start from a social science perspective, explicating the socially important aspects of individual agency and social structure. Instead of a reporting protocol, they present a framework through which to increase the transparency of agent-based models in the social sciences by linking "conceptual elements of [agent-based models] and some of the important dimensions of social reality" (Dilaver and Gilbert, 2023, 4.7). They also provide a helpful table as a guide to describing a model.

14.6.2 Minimal Requirements

Regardless of whether you follow one of the protocols, there is certain information that must be reported about any model used for research. You can use the Table of broad design questions (Table 2.1) as a starting point. This would be supplemented with decisions that you have made along the way that a reader needs to know to understand the model design and how you have used it.

The following is a list of aspects of the research that should also be reported in any publication.

1. Theoretical embedding: Your model needs embedding into the theoretical or policy background it sits in. A review of the relevant literature of research on the general topic and some of the specifics of the model, e.g. game theory if that is the foundation of your model.
2. Model assumptions: You should always clearly state the assumptions the model makes. This includes decisions of abstraction, idealisation, and operationalisation.
3. Research hypothesis: State clearly the research question(s) you are trying to tackle and the hypotheses underlying your research.
4. Justification for the approach: You should justify why you took an agent-based modelling approach, rather than other kinds of models or even a statistical analysis.
5. Model outline: Describe the model. This does not need to include implementation detail but should have enough detail so the reader understands the relevant agent characteristics and how the mechanisms are embodied by the model.
6. Experimental design: Describe the experiments you have conducted, including the parameters, the values, and how many runs.
7. Model outputs: Report on the model outputs. Produce plots for what the model allows us to measure.
8. Model robustness: Report on how robust your modelling results are. At best this should be a full sensitivity analysis, but sometimes it is enough to check for the robustness of a set of configurations and results.
9. Conclusions: Draw a set of sensible conclusions from your model. Make sure you do not overstate what the model allows us to believe. Also make sure you relate your conclusions back to the initial research questions and hypotheses.
10. Optional Appendices:
    (a) Detailed model description (including implementation details), potentially following the ODD format if that is appropriate.

(b) The program code or a link to the published model.

(c) Tables of additional outputs and results.

(d) Dataset of simulation output.

Be prepared to get reviews that will question the overall approach you have taken in choosing agent-based modelling. Although an increasingly accepted method in general, many journals, in particular disciplinary journals, will ask for justification why you step out of existing disciplinary methodological paradigms. Other review challenges to be prepared for include: how you have validated your model, justification of the assumptions you make in the design, including why you left things out. Keeping these potential reviewer challenges in mind while writing your article will make your life easier.

14.7 Summary

In this chapter, we have discussed a range of things that need doing once the model is built. First of all, the model needs to be verified, checked that it does what it is supposed to do. The tutorial showed you many ways in which you 'do' verification while building the model; incremental changes that are each tested to see whether it works the way it was intended to work. We then looked at how an agent-based model relates to real-world data through calibration and validation. Validation is an important, yet contentious, aspect of agent-based modelling. It is important because the model is supposed to represent the relevant features, relationships, and processes of (part of) the real world, and contentious because of the great variety of agent-based models and their different needs for empirical validation. Pattern Oriented Modelling is often used as a way to tailor the patterns and processes to the complexity and realism of the model.

Finally, we discussed communicating your model. Again, the variety of models makes it difficult to unify communication. The ODD protocol is by far the most accepted reporting tool, but that does not mean it is appropriate for all models. It was designed for transparency and communication with other modellers to describe all the details necessary to reproduce the model. That makes it cumbersome for more abstract models and lengthy, with excessive detail not relevant for most readers that may obscure the key process descriptions necessary to understand the model.

In Section 2.5.1, we discussed that agent-based modelling is about sensemaking, finding possible explanations of the phenomena that surround us. It is helpful to keep this qualitative approach to agent-based modelling in mind when reporting and communicating the model. It will help you tell the "justified stories" the model generates rather than focusing too much on replicating a particular outcome (Badham et al., 2021; Elsenbroich and Badham, 2023).

What is important is to be transparent in your communication about modelling choices you made and limitations of the model. Table 2.1 can help with this as it walks you through many of the choices you have to make. If you record them while building your model, you will have a good basis from which to report later on. Similarly, while building the model, annotate your code. The tutorial teaches you good coding practices, of which commenting is one important aspect. Models, including the code, should be published so that other people have access to be able to check your claims.

Questions

1. What has been your most common mistake while building the model in the tutorial?
2. Name one NetLogo feature that helps you verify your model.
3. List three ways in which empirical data is used within agent-based modelling.
4. What is calibration of an agent-based model?
5. Discuss why standardising validation approaches for agent-based modelling is difficult.
6. Take a model from the NetLogo Model Library (best go for the Social Science Section) and try to do an ODD and/or a vODD visualisation for it.

15

Being an Agent-Based Modeller

Congratulations — you are now an agent-based modeller! You have built your first agent-based model and thought about all kinds of aspects of modelling, including conceptualisation, operationalisation, experimentation, validation, and interpretation. You might not feel like a full-blown agent-based modeller just yet, but you have the skills and knowledge that you need to build your own model. Hopefully this will be with other colleagues who understand agent-based modelling, and from whom you can learn, but that might not always be the case.

15.1 Takeaway Messages for Agent-based Modelling

Even if you never build another agent-based model, you might see the world in a slightly different way, being aware of the importance of individuals and their interactions, of emergence, and of the need to simplify things in order to understand them, but to not oversimplify and distort. Most importantly, you should be sensitised to search for *processes*, *emergence*, and *complexity*. Agent-based modelling is a methodology that helps us think about these three things in a structured and transparent way. By modelling societies as complex systems composed of interdependent and interacting components, such as individuals, institutions, and resources, agent-based models explain system properties like non-linearity, emergence, and dynamic change that are impossible to understand using reductive methods that isolate components from their context and difficult to think through without the help of a computer.

Agent-based modelling is based on the idea of Generative Social Science, where the models serve as tools to explain social phenomena by reconstructing the processes that give rise to them. Rather than simply describing observed outcomes, agent-based modelling allows researchers to experiment with different assumptions and mechanisms to see how complex outcomes can emerge. Ultimately, agent-based modelling supports a non-reductivist, process-focused approach to understanding social complexity, offering a powerful lens through which to examine and theorise about the dynamic patterns and structures of human behaviour and interaction. This view of agent-based modelling makes it a methodology rather than a method. It is not merely a simulation technique but a broad, theory-informed framework that guides how we conceptualise, design, and interpret the world.

This makes agent-based models particularly well-suited to capture the dynamics of social phenomena, from segregation and opinion formation to economic exchange and collective behaviour. As computational power and data availability have grown, so too has the complexity and realism of agent-based models, which are now used to study real-world dynamics in areas such as public health, crime, land use, and technology adoption. But whether simple or complex, abstract or more realistic, at the heart of agent-based models is the concept of autonomous agents and the processes through which they interact with each other and their environment. We discussed *agent-centric thinking* as the crux of agent-based modelling, identifying who the agents are, what they know and can perceive, what rules govern an individual's behaviour, how they interact, and how those interactions give rise to broader system-level outcomes (see Section 2.1.3).

Building an agent-based model is an iterative and reflective process. It involves defining agents, their environment, and the rules of interaction, but also demands careful attention to abstraction, causal mechanisms, data integration, and model outputs. As modelling progresses, you will often revisit and refine earlier decisions in response to new insights or data, an essential part of rigorous model development rather than a sign of inconsistency. This iterative engagement not only strengthens the model itself but also deepens your understanding of the system under study. This methodological perspective aligns agent-based modelling more closely with qualitative approaches than with traditional quantitative methods, given the active role of the researcher in model design, abstraction, representation, and interpretation. Agent-based modelling is a sense-making activity, helping us understand why we observe particular phenomena.

Finally we would like to reiterate the importance of the communication of agent-based models. On the documentation side, the ODD protocol provides a widely adopted template for describing model purpose, entities, processes, and assumptions, though it may be too rigid for highly abstract or theoretical models. Recent proposals, such as the framework by Dilaver and Gilbert (2023), aim to broaden documentation standards to better accommodate diverse modelling approaches.

When it comes to interpreting results, modellers must resist the temptation to overstate their model's predictive power. A conformity agent-based model, for instance, cannot legitimately predict an election outcome simply by adding a "charismatic leader" parameter or running on a geo-specific synthetic population. Conclusions should stay aligned with the model's original intent, be it explanation, scenario exploration, or hypothesis testing, and avoid "suggestivism", where uncertainty is obscured by vague or overconfident claims (Edmonds, 2022). By combining rigorous validation, thoughtful application of new data sources, and clear, honest communication, the agent-based modelling community can continue to enhance both the credibility and the insightfulness of agent-based research.

15.2 Current Developments in Agent-Based Modelling

Agent-based models have come a long way in the past 20 years. What began as simple, low-dimensional "toy" models has matured into rich, multi-dimensional simulations grounded in complex understandings of social phenomena with highly realistic agent behaviours and tackling some of the most entrenched problems facing us today (Castro et al., 2020). Agent-based models support research across many disciplines, with the most developed empirical models in urban planning, land use, and environmental management (Railsback and Grimm, 2011); they are also making inroads into informing policy (Gilbert et al., 2018). Their visibility soared during the COVID-19 pandemic, when they helped estimate infection trajectories under various non-pharmaceutical interventions, gauge health-care capacity needs, and guide public-health planning (Giabbanelli et al., 2021). Of course, the reason for their prominence as models for non-pharmaceutical interventions is that these interventions all changed the interaction patterns of individuals. More recently, fields such as public health and criminology have embraced agent-based modelling to tackle entrenched, complex problems such as health inequalities (Boyd et al., 2022) or obesity (Giabbanelli et al., 2021) or to prototype interventions (Birks et al., 2025).

Despite their growing uptake, agent-based models still wrestle with two major challenges. The first is validation. Unlike statistical models that directly fit data, agent-based models reconstruct the underlying processes thought to generate observed phenomena; yet multiple equally plausible processes can yield the same macro-level patterns (the problem of equifinality; see Chapter 14). As a result, establishing that a model is truly representative of reality remains a matter of debate. The second enduring question concerns prediction. Some argue that agent-based models should focus on explanation and exploration, illuminating "what if?" scenarios rather than making point forecasts (Edmonds, 2017). Others insist that we can reduce uncertainty in prediction by using more and better data to calibrate our agent-based models (Vu et al., 2019). You can find a comprehensive discussion of positions on prediction in Elsenbroich and Polhill (2023).

Over the last decade, agent-based modelling has become much more data-focused, or even data-driven (e.g. Rosés et al., 2021). Emerging data streams, such as real-time sensors tracking mobility, wearable health monitors, and detailed transaction records, offer exciting new inputs for urban and transportation agent-based models (Heppenstall et al., 2021). These new data help us understand new facets of societies and can be useful for agent-based modelling as they are temporally explicit data streams, a kind of data often wanted for agent-based models. A second data focus is the integration of agent-based models with synthetic populations or "digital twins" (Jiang et al., 2022). These populations, generated by combining census data with surveys, geographical data, or administrative records, allow us to scale up models to realistic populations. This can help us understand effects of modelled dynamics for specific populations, for example differences of COVID-19 outcomes in populations with different age profiles due to the age sensitivity of COVID-19.

Whilst all these new developments are exciting, it remains unclear how well these high-resolution data can capture the richer decision-making and social interactions that drive emergent system outcomes and thus need replication in agent-based models. Although big data can enhance calibration and validation, blindly extracting behavioural rules risks oversimplifying the diversity of processes that give rise to complex social patterns. Implementing agent-based models on synthetic populations or fine-grained GIS data risks overreaching their empirical basis. A model with highly abstract opinion dynamics rules, implemented on a detailed synthetic population and GIS mapping is still only a highly abstract and simple model. A Susceptible-Infected-Recovered (SIR) epidemic model that has planes flying between countries with timetable informed schedules enriches the representation of mixing patterns while travelling but remains limited by lack of knowledge about mixing elsewhere, differential susceptibility, and other unknowns. The crux for agent-based models lies in the empirical foundation of the relevant processes, not the static data.

We would thus like to caution against an over-reliance on large data sets for building agent-based models. As mentioned several times before, an agent-based model is not a summary model of a dataset, akin to a statistical model. An agent-based model is instead an explanation of why we find the patterns that we can extract from the dataset. It can use data as the 'traces of processes', system states that let us infer what must have been occurring to get from one state to the next. So while new data streams and synthetic populations are useful for agent-based modelling, they do not, in our opinion, change the hard graft that needs to be done by the modeller to make decisions about model boundaries, understand and select micro-behaviours, and abstract and idealise to operationalise our representation of the real world into a computational model. The quality of the representation in an agent-based model is limited by the lowest fidelity piece of the representation.

A second development, connected but not identical with data-driven agent-based modelling is the surge of Large Language Models and Generative AI, which promise new ways of dealing with large, dynamic datasets and can be used to extract patterns and rules across data or even to *be* the agents in an agent-based model (Gao et al., 2024). Large Language Models have the potential to transform traditional agent-based modelling by enabling AI-generated, human-like, intelligent agents integrating real-time data flows into behaviour choices. As of now, in 2026, this is a recent but rapidly expanding development.

We want to caution against a wholesale adoption of data- and AI-driven behaviour extractions, except maybe for very simple behaviours, such as pedestrians in a city. The reason for this is that we see models as 'thinking tools'. They help us to think about the unfolding of events, simplifying the complexity of the real world enough for us to follow but not so much to distort events. Replacing rules of behaviour by finely calibrated probabilities of action might allow us a closer fit of the model to real-world data but it fundamentally changes the usability and purpose of the model as it is now a mere projection tool (Elsenbroich and Badham, 2023). The process of conceptualising, designing, and building the model is as much part of understanding as the analysis of the final product. Outsourcing this process and excluding theory from the model design diminishes the usefulness of the model, potentially eradicating it completely.

15.3 Limitations of Agent-based Modelling

Throughout the book we have been singing the praises of agent-based modelling. Of course it is not the universal solution to all research problems; it is a very rich methodology to understand processes and complexity resulting from interactions over time.

There are some common criticisms of agent-based modelling. The first one is that agent-based models tell us nothing new about the world. You simply "get out what you put in". Of course it is true that models rely on their inputs. Despite that, agent-based models still enhance our understanding of systems and their behaviours. The iterated, stochastic, sometimes random interactions of system components are impossible to run in our heads. Agent-based models are the tool that helps us think about the consequences of these interactions.

Having said this, agent-based models can be overdesigned and overfitted to make specific points. When building an agent-based model it is important to ensure that the patterns the model is trying to explain are not programmed into the model from the start. You cannot explain health inequalities by starting with agents that have unequal health. Instead, those inequalities should arise from other features and processes of the model. The agent-centric perspective helps you make decisions about what needs 'putting into' an agent-based model. Focusing in on the micro-level can help to not overload the model with the outcomes one is expecting. The model allows those outcomes to be generated by the mechanisms.

It is, however, also important to acknowledge that an agent-based model is the result of many decisions (see Section 2.3), of abstractions and idealisations, of assumptions and potential biases.

It is necessary to 'get our hands dirty' as researchers if we want to understand the world. As Tolk puts it:

> Modeling is the task-driven, purposeful simplification and abstraction of a perception of reality that is shaped by physical, ethical and cognitive constraints. (Tolk, 2017, p.47)

That means you need to be transparent about your modelling choices and able to defend them.

A second common criticism is that different models can produce the same behaviour. Equifinality is indeed a problem but is not specific to agent-based modelling. In fact, equifinality is a feature of complex systems, where different causal paths can lead to the same outcome (Luhmann, 1995). As a feature of agent-based models themselves, equifinality can be tackled by the different kinds of validation we discussed in Section 14.4. Just focusing on outcomes leaves models open to equifinality but additional validation of the inputs and processes supports the model being the right model of the target system. It is also important to stress that agent-based models do not exist in a vacuum. Whilst a single model may not be able to determine its own truth, comparing different models and integrating models into increasing levels of system-knowledge resulting from other methods allows for more robust conclusions.

A third often advanced critique is that agent-based models are just too complicated to explain anything. The world itself is complex, and models must reflect this. Models should be "as simple as possible, but no simpler", that Einstein aphorism yet again! Agent-based models *are* the representation of the relevant features, relationships, and processes to help us understand the world better. It is less important to build simple models and more important to build models that are transparent, tractable, and interpretable in the right way.

As a final limitation, we would like to discuss what kinds of questions an agent-based model is able to answer. It is important to use agent-based models for the right kinds of questions and be clear what questions these models can and cannot answer. Coming back to the start of this book, a model is a representation of the relevant features and relationships of the target system. The first distinction to be made is between empirical questions and modelling questions. An agent-based model, however good it is, will not be able to answer empirical questions like how do people understand, feel about, or experience something. To answer those questions you need to ask people, through surveys, interviews, or focus groups. A model will use this empirical knowledge to help us understand how people's feelings, understandings, and interpretations might lead to changed behaviour that will produce certain societal outcomes. A model will not be able to answer the question of how people will change their behaviour; the model will help us think through what changes IF people change their behaviours in certain ways, and it will particularly help us to think through these questions where there are social interactions so that changes in one person's behaviour brings about changes in others. For example, an agent-based model will not answer the question of how people will change their eating habits if all fast food advertising is removed. It will, however, answer the question of how much less fast food might be consumed IF people stop buying fast food unless they see advertising for it.

15.4 Telling Justified Stories with Agent-based Models

Finally, we want to clarify our position on agent-based models. Whilst we tried to keep the book relatively descriptive regarding agent-based modelling, our introduction of epistemology, ontology, and methodology paved the way to talk more broadly about agent-based modelling. We argue that agent-based modelling is not a technical tool to add to your data analysis skills. In fact, we do not think that it is a data analysis tool at all. Agent-based modelling is a methodology that helps us think about change in a complex world. It does that by helping us to tell "justified stories" about the world. Justified stories are internally coherent, theory-driven scenarios validated against real-world data. They do not predict a specific future but help decision-makers grasp how different futures could plausibly emerge, enhancing understanding rather than delivering forecasts (Elsenbroich and Badham, 2023).

In a complex and dynamic world, we often need to make critical decisions with incomplete information. Traditional predictive models, such as statistical and machine learning models, rely heavily on past data and assume the future will resemble the past. However, this assumption breaks down in complex systems where small changes can lead to large, unpredictable effects. Agent-based modelling is a structured and transparent way to explore future possibilities. Models combine theory (causal mechanisms and assumptions) with data (empirical calibration) to allow us to tell justified stories about how the future might unfold.

The shift away from prediction or even forecasting towards the telling of justified stories brings with it a different approach to building and using agent-based models, hence our stress on the importance of understanding micro-processes, being transparent about modelling decisions and validating input as well as modelling outputs against data. In terms of working with agent-based models, the justified stories approach will fundamentally change the desired outputs of agent-based models. Rather than focusing on particular system states to be reached (a focus necessary for prediction), the justified stories approach focuses on the paths the simulation model outlines. Rather than focusing on averages across runs, the justified stories approach looks at the spread, at outliers that suggest tipping points and path dependencies.

Agent-based models range from abstract conceptual models to highly detailed, data-driven simulations of real-world systems. Regardless of their form, they aim to capture the essential features, relationships, and processes of the system they represent. But a model is not an oracle: it is a structured way to think about the world. It allows us to explore both expected and unexpected consequences, helping researchers and policymakers make better-informed decisions. Ultimately, agent-based models serve as powerful tools for understanding complexity, rather than providing definitive answers.

Bibliography

Ajzen, I. (1991). The theory of planned behavior. *Organizational Behavior and Human Decision Processes 50*(2), 179–211.

Ajzen, I. (2011). The theory of planned behaviour: Reactions and reflections.

Anshuka, A., F. F. van Ogtrop, D. Sanderson, and S. Z. Leao (2022). A systematic review of agent-based model for flood risk management and assessment using the ODD protocol. *Natural Hazards 112*(3), 2739–2771.

Anzola, D., P. Barbrook-Johnson, and N. Gilbert (2022). The ethics of agent-based social simulation. *Journal of Artificial Societies and Social Simulation 25*(4), 1.

Aristotle (2011). *Aristotle's Metaphysics.* Green Lion Press. Translated by Joe Sachs.

Axelrod, R. (1980). Effective choice in the prisoner's dilemma. *Journal of Conflict Resolution 24*(1), 3–25.

Axelrod, R. (1997). The complexity of cooperation: Agent-based models of competition and collaboration. In *The complexity of cooperation.* Princeton University Press.

Axelrod, R. and W. D. Hamilton (1981). The evolution of cooperation. *Science 211*(4489), 1390–1396.

Badham, J., P. Barbrook-Johnson, C. Caiado, and B. Castellani (2021). Justified stories with agent-based modelling for local COVID-19 planning. *Journal of Artificial Societies and Social Simulation 24*(1), 8.

Badham, J., S. Elsawah, J. H. Guillaume, S. H. Hamilton, R. J. Hunt, A. J. Jakeman, S. A. Pierce, V. O. Snow, M. Babbar-Sebens, B. Fu, P. Gober, M. C. Hill, T. Iwanaga, D. P. Loucks, W. S. Merritt, S. D. Peckham, A. K. Richmond, F. Zare, D. Ames, and G. Bammer (2019). Effective modeling for integrated water resource management: A guide to contextual practices by phases and steps and future opportunities. *Environmental Modelling & Software 116*, 40–56.

Badham, J., C. Jansen, N. Shardlow, and T. French (2017). Calibrating with multiple criteria: A demonstration of dominance. *Journal of Artificial Societies and Social Simulation 20*(2), 11.

Balke, T. and N. Gilbert (2014). How do agents make decisions? A survey. *Journal of Artificial Societies and Social Simulation 17*(4), 13.

Barbrook-Johnson, P., J. Badham, and N. Gilbert (2017). Uses of Agent-Based Modeling for Health Communication: the TELL ME Case Study. *Health Communication 32*(8), 939–944.

Barlas, Y. and S. Carpenter (1990). Philosophical roots of model validation: two paradigms. *System Dynamics Review 6*(2), 148–166.

Barton, A. H. (1968). Survey research and macro-methodology. *American Behavioral Scientist 12*(2), 1–9.

Bass, F. M. (1969). A new product growth for model consumer durables. *Management Science 15*(5), 215–227.

Belfrage, M., E. Johansson, F. Lorig, and P. Davidsson (2024). [In]Credible models – verification, validation & accreditation of agent-based models to support policy-making. *Journal of Artificial Societies and Social Simulation 27*(4), 4.

Birks, D., E. R. Groff, and N. Malleson (2025). Agent-based modeling in criminology. *Annual Review of Criminology 8*(1), 75–95.

Borgonovo, E., M. Pangallo, J. Rivkin, L. Rizzo, and N. Siggelkow (2022). Sensitivity analysis of agent-based models: a new protocol. *Computational and Mathematical Organization Theory 28*(1), 52–94.

Boyd, J., R. Wilson, C. Elsenbroich, A. Heppenstall, and P. Meier (2022). Agent-based modelling of health inequalities following the complexity turn in public health: a systematic review. *International Journal of Environmental Research and Public Health 19*(24), 16807.

Bratman, M. (1984). Two faces of intention. *The Philosophical Review 93*(3), 375–405.

Cartwright, N. and J. Hardie (2012). *Evidence-based policy: A practical guide to doing it better.* Oxford University Press.

Castellani, B., R. Rajaram, J. G. Buckwalter, M. Ball, and F. Hafferty (2015). *Place and health as complex systems: A case study and empirical test.* Springer.

Castro, J., S. Drews, F. Exadaktylos, J. Foramitti, F. Klein, T. Konc, I. Savin, and J. van Den Bergh (2020). A review of agent-based modeling of climate-energy policy. *Wiley Interdisciplinary Reviews: Climate Change 11*(4), e647.

Center for Connected Learning & Computer-Based Modeling (2021). Common NetLogo error messages. `https://ccl.northwestern.edu/netlogo/bind/article/common-netlogo-error-messages.html`.

Chapuis, K., P. Taillandier, and A. Drogoul (2022). Generation of synthetic populations in social simulations: a review of methods and practices. *Journal of Artificial Societies and Social Simulation 25*(2), 6.

Chattoe-Brown, E. (2009). Gin and tonic or oil and water? can we integrate research methods rather than just "mixing" them? a case study of social mobility research. *Draft Paper, Department of Sociology, University of Leicester*.

Chattoe-Brown, E. (2013). Why sociology should use agent based modelling. *Sociological Research Online 18*(3), 3.

Chattoe-Brown, E., N. Gilbert, D. A. Robertson, and C. Watts (2021). Reproduction as a means of evaluating policy models: A case study of a covid-19 simulation. *medRxiv*.

Collins, A., M. Koehler, and C. Lynch (2024). Methods that support the validation of agent-based models: An overview and discussion. *Journal of Artificial Societies and Social Simulation 27*(1), 11.

Crooks, A. (2016). The use of agent-based modeling for studying the social and physical environment of cities. In *Complexity and Planning*, pp. 385–408. Routledge.

Crooks, A., N. Malleson, E. Manley, and A. Heppenstall (2018). *Agent-Based Modelling and Geographical Information Systems*. SAGE Publications Ltd.

Deffuant, G., F. Amblard, G. Weisbuch, and T. Faure (2002). How can extremism prevail? A study based on the relative agreement interaction model. *Journal of Artificial Societies and Social Simulation 5*(4), 1.

Diekmann, O. and J. A. P. Heesterbeek (2000). *Mathematical Epidemiology of Infectious Diseases: Model Building, Analysis and Interpretation*. Wiley.

Dignum, F. (2021). *Social Simulation for a Crisis*. Springer.

Dilaver, O. and N. Gilbert (2023). Unpacking a black box: a conceptual anatomy framework for agent-based social simulation models. *Journal of Artificial Societies and Social Simulation 26*(1), 4.

Edmonds, B. (2017). *Different modelling purposes*, pp. 39–58. Springer.

Edmonds, B. (2022, February). The poverty of suggestivism-the dangers of "suggests that" modelling. *Review of Artificial Societies and Social Simulation*.

Edmonds, B. (2023). A declaration of social simulator rights. In *Conference of the European Social Simulation Association*, pp. 283–289. Springer.

Edmonds, B. and S. Moss (2004). From KISS to KIDS–an 'anti-simplistic' modelling approach. In *International workshop on multi-agent systems and agent-based simulation*, pp. 130–144. Springer.

Elsenbroich, C. (2017). The Addio Pizzo movement: exploring social change using agent-based modelling. *Trends in Organized Crime 20*, 120–138.

Elsenbroich, C. and J. Badham (2020, April). Focussing on our strengths. *Review of Artificial Societies and Social Simulation*.

Elsenbroich, C. and J. Badham (2023). Negotiating a future that is not like the past. *International Journal of Social Research Methodology 26*(2), 207–213.

Elsenbroich, C. and N. Gilbert (2014). *Modelling Norms*. Springer.

Elsenbroich, C. and J. G. Polhill (2023). Agent-based modelling as a method for prediction in complex social systems. *International Journal of Social Research Methodology 26*(2), 133–142.

Elster, J. (1989). *Nuts and Bolts for the Social Sciences*. Cambridge University Press.

Epstein, J. M. (2006). *Generative Social Science: Studies in agent-based computational modeling*. Princeton, NJ: Princeton University Press.

Epstein, J. M. (2008). Why model? *Journal of Artificial Societies and Social Simulation 11*(4), 12.

Epstein, J. M. and R. Axtell (1996). *Growing Artificial Societies: Social science from the bottom up*. Brookings Institution Press.

Farley, D. (2022). *Modern software engineering.* Boston, MA: Addison Wesley.

Ferguson, N. M., D. Laydon, G. Nedjati-Gilani, N. Imai, K. Ainslie, M. Baguelin, S. Bhatia, A. Boonyasiri, Z. Cucunubá, G. Cuomo-Dannenburg, et al. (2020). *Report 9: Impact of non-pharmaceutical interventions (NPIs) to reduce COVID19 mortality and healthcare demand,* Volume 16. Imperial College London London.

Fuster, A., P. Goldsmith-Pinkham, T. Ramadorai, and A. Walther (2022). Predictably unequal? the effects of machine learning on credit markets. *The Journal of Finance 77*(1), 5–47.

Gallagher, C. A., M. Chudzinska, A. Larsen-Gray, C. J. Pollock, S. N. Sells, P. J. C. White, and U. Berger (2021). From theory to practice in pattern-oriented modelling: identifying and using empirical patterns in predictive models. *Biological Reviews 96*(5), 1868–1888.

Gamal, Y., C. Elsenbroich, N. Gilbert, A. Heppenstall, and K. Zia (2024). A behavioural agent-based model for housing markets: impact of financial shocks in UK. *Journal of Artificial Societies and Social Simulation 27*(4), 5.

Gao, C., X. Lan, N. Li, Y. Yuan, J. Ding, Z. Zhou, F. Xu, and Y. Li (2024). Large language models empowered agent-based modeling and simulation: A survey and perspectives. *Humanities and Social Sciences Communications 11*(1), 1–24.

Giabbanelli, P. J., J. Badham, B. Castellani, H. Kavak, V. Mago, A. Negahban, and S. Swarup (2021). Opportunities and challenges in developing COVID-19 simulation models: Lessons from six funded projects. In *2021 Annual Modeling and Simulation Conference (ANNSIM).* IEEE.

Giabbanelli, P. J., B. Tison, and J. Keith (2021). The application of modeling and simulation to public health: Assessing the quality of agent-based models for obesity. *Simulation Modelling Practice and Theory 108*(102268), 102268.

Giardini, F., R. Conte, and M. Paolucci (2013). *Reputation*, pp. 365–399. Springer.

Giddens, A. (1984). *The Constitution of Society: Outline of the theory of structuration.* Polity.

Gilbert, N. (2008). *Agent-Based Models.* Quantitative Applications in the Social Sciences. Los Angeles: Sage Publications.

Gilbert, N., P. Ahrweiler, P. Barbrook-Johnson, K. P. Narasimhan, and H. Wilkinson (2018). Computational modelling of public policy: Reflections on practice. *Journal of Artificial Societies and Social Simulation 21*(1), 14.

Gilbert, N., P. Ahrweiler, A. Pyka, et al. (2014). *Simulating Knowledge Dynamics in Innovation Networks.* Springer.

Gilbert, N., E. Chattoe-Brown, C. Watts, and D. Robertson (2021). Why we need more data before the next pandemic. *Sociologica 15*(3), 125–143.

Gilbert, N., J. C. Hawksworth, and P. A. Swinney (2009). An agent-based model of the English housing market. In *AAAI spring symposium: technosocial predictive analytics*, pp. 30–35.

Greenberger, M., M. A. Censon, and B. L. Crissey (1976). *Models in the policy process: Public decision making in the computer era.* Russell Sage Foundation.

Grimm, V., U. Berger, F. Bastiansen, S. Eliassen, V. Ginot, J. Giske, J. Goss-Custard, T. Grand, S. K. Heinz, G. Huse, et al. (2006). A standard protocol for describing individual-based and agent-based models. *Ecological modelling 198*(1-2), 115–126.

Grimm, V., U. Berger, D. L. DeAngelis, J. G. Polhill, J. Giske, and S. F. Railsback (2010). The ODD protocol: A review and first update. *Ecological Modelling 221*(23), 2760–2768.

Grimm, V., K. Frank, F. Jeltsch, R. Brandl, J. Uchmański, and C. Wissel (1996). Pattern-oriented modelling in population ecology. *Science of the Total Environment 183*(1), 151–166.

Grimm, V. and S. F. Railsback (2012). Pattern-oriented modelling: a 'multi-scope' for predictive systems ecology. *Philosophical Transactions of the Royal Society B: Biological Sciences 367*(1586), 298–310.

Grimm, V., S. F. Railsback, C. E. Vincenot, U. Berger, C. Gallagher, D. L. DeAngelis, B. Edmonds, J. Ge, J. Giske, J. Groeneveld, et al. (2020). The ODD protocol for describing agent-based and other simulation models: A second update to improve clarity, replication, and structural realism. *Journal of Artificial Societies and Social Simulation 23*(2), 7.

Grimm, V., E. Revilla, U. Berger, F. Jeltsch, W. M. Mooij, S. F. Railsback, H.-H. Thulke, J. Weiner, T. Wiegand, and D. L. DeAngelis (2005). Pattern-oriented modeling of agent-based complex systems: Lessons from ecology. *Science 310*(5750), 987–991.

Grow, A. and A. Flache (2019). *Agent-based computational models of reputation and status dynamics*, pp. 230. Oxford University Press Oxford.

Hailegiorgis, A., A. Crooks, and C. Cioffi-Revilla (2018). An agent-based model of rural households' adaptation to climate change. *Journal of Artificial Societies and Social Simulation 21*(4), 4.

Hamill, L. and N. Gilbert (2015). *Agent-based Modelling in Economics*. John Wiley & Sons.

Hammond, R. A. and R. Axelrod (2006). The evolution of ethnocentrism. *Journal of Conflict Resolution 50*(6), 926–936.

Hannappel, M., K. G. Troitzsch, and S. Bauschke (2012). Demographic and educational projections. building an event-oriented microsimulation model with CoMICS II. In *ECMS*, pp. 613–618.

Hartmann, S. (1996). The world as a process: Simulations in the natural and social sciences. In R. Hegselmann, U. Mueller, and K. G. Troitzsch (Eds.), *Modelling and Simulation in the Social Sciences from the Philosophy of Science Point of View*, pp. 77–100. Springer.

Hegselmann, R. (2012). Thomas C. Schelling and the computer: Some notes on Schelling's essay "on letting a computer help with the work". *Journal of Artificial Societies and Social Simulation 15*(4), 9.

Hegselmann, R. (2017). Thomas C. Schelling and James M. Sakoda: The intellectual, technical, and social history of a model. *Journal of Artificial Societies and Social Simulation 20*(3), 15.

Hegselmann, R. and U. Krause (2002). Opinion dynamics and bounded confidence: models, analysis and simulation. *Journal of Artificial Societies and Social Simulation 5*(3), 2.

Hempel, C. G. (2014). The theoretician's dilemma a study in the logic of theory construction. In *The Nature of Scientific Theory*, pp. 99–160. Routledge.

Heppenstall, A., A. Crooks, N. Malleson, E. Manley, J. Ge, and M. Batty (2021). Future developments in geographical agent-based models: Challenges and opportunities. *Geographical Analysis 53*(1), 76–91.

Hume, D. (1894). *An Enquiry Concerning the Human Understanding: And an enquiry concerning the principles of morals.* Clarendon Press.

Jager, W. (2017). Enhancing the realism of simulation (EROS): On implementing and developing psychological theory in social simulation. *Jasss-The journal of artificial societies and social simulation 20*(3), 14.

Janssen, M. (2020). Sugarscape with spice. `https://www.comses.net/codebases/6f72dfc9-e8c5-4d9c-b925-7338059573c3/releases/1.1.0/`.

Janssen, M. A., D. DeCaro, and A. Lee (2022). An agent-based model of the interaction between inequality, trust, and communication in common pool experiments. *Journal of Artificial Societies and Social Simulation 25*(4), 3.

Jiang, N., A. T. Crooks, H. Kavak, A. Burger, and W. G. Kennedy (2022). A method to create a synthetic population with social networks for geographically-explicit agent-based models. *Computational Urban Science 2*(1), 7.

Keane, C. (2013). *Modeling Behavior in Complex Public Health Systems: Simulation and Games for Action and Evaluation.* Springer.

Kermack, W. O. and A. G. McKendrick (1927). A contribution to the mathematical theory of epidemics - I. *Proceedings of the Royal Society A: Mathematical, Physical and Engineering Sciences 115A*(772), 700–721.

Kincaid, H. (2014). *Dead ends and live issues in the individualism-holism debate*, pp. 139–152. Springer.

Klein, D. and J. Marx (2018). Generalized trust in the mirror. an agent-based model on the dynamics of trust. *Historical Social Research/Historische Sozialforschung 43*(1), 234–258.

Knuth, D. E. (1984). Literate programming. *The Computer Journal 27*(2), 97–111.

Kuhn, T. S. (1970). *The Structure of Scientific Revolutions*, Volume 111. Chicago University of Chicago Press.

Lave, C. A. and J. G. March (1993). *An Introduction to Models in the Social Sciences.* Lanham, MD: University Press of America.

Leslie, D. (2020). Understanding bias in facial recognition technologies. *arXiv preprint arXiv:2010.07023*.

Lettieri, N. and D. Parisi (2013). Neminem laedere. An evolutionary agent-based model of the interplay between punishment and damaging behaviours. *Artificial Intelligence and Law 21*, 425–453.

Li, J. and U. Wilensky (2009). NetLogo Sugarscape 3 Wealth Distribution Model.

Lotka, A. J. (1925). *Elements of Physical Biology*. Baltimore, MD: Williams & Wilkins Co.

Luhmann, N. (1995). *Social Systems*. Stanford University Press.

Macal, C. M. and M. J. North (2010). Tutorial on agent-based modelling and simulation. *Journal of Simulation 4*(3), 151–162.

Malleson, N., M. Birkin, D. Birks, J. Ge, A. Heppenstall, E. Manley, J. McCulloch, and P. Ternes (2022). Agent-based modelling for urban analytics: State of the art and challenges. *AI Communications 35*(4), 393–406.

Manzo, G. (2023). Agent-based models and methodological individualism: Are they fundamentally linked? In *The Palgrave Handbook of Methodological Individualism: Volume I*, pp. 681–710. Springer.

Mastroeni, L., P. Vellucci, and M. Naldi (2019). Agent-based models for opinion formation: A bibliographic survey. *IEEE Access 7*, 58836–58848.

McCulloch, J., J. Ge, J. A. Ward, A. Heppenstall, J. G. Polhill, and N. Malleson (2022). Calibrating agent-based models using uncertainty quantification methods. *Journal of Artificial Societies and Social Simulation 25*(2), 1.

McKay, M. D., R. J. Beckman, and W. J. Conover (1979). A comparison of three methods for selecting values of input variables in the analysis of output from a computer code. *Technometrics 21*(2), 239.

Miller, J. H. and S. E. Page (2007). *Complex Adaptive Systems*. University Press Group Ltd.

Morris, M. D. (1991). Factorial sampling plans for preliminary computational experiments. *Technometrics 33*(2), 161–174.

Moss, S. (2007). Alternative approaches to the empirical validation of agent-based models. *Journal of Artificial Societies and Social Simulation 11*(1), 5.

Mossong, J., N. Hens, M. Jit, P. Beutels, K. Auranen, R. Mikolajczyk, M. Massari, S. Salmaso, G. S. Tomba, J. Wallinga, et al. (2008). Social contacts and mixing patterns relevant to the spread of infectious diseases. *PLoS Medicine 5*(3), e74.

Müller, B., S. Balbi, C. M. Buchmann, L. De Sousa, G. Dressler, J. Groeneveld, C. J. Klassert, Q. B. Le, J. D. Millington, H. Nolzen, et al. (2014). Standardised and transparent model descriptions for agent-based models: Current status and prospects. *Environmental Modelling & Software 55*, 156–163.

Narasimhan, K., T. Roberts, M. Xenitidou, and N. Gilbert (2017). Using ABM to clarify and refine social practice theory. In *Advances in Social Simulation 2015*, pp. 307–319. Springer.

Ncube, N. N., C. Goodfellow, M. McCann, J. Riddell, J. Allen, E. Long, C. White, J. Cameron, L. Barsalou, and C. Elsenbroich (2024). A SOCial sITuational Systems approach to participatory qualitative inquiry: Using remote walking interviews and pupil-teacher workshops to identify social situations related to adolescent mental health. *OSF preprints*.

Newman, M. (2010). *Networks: An Introduction*. Oxford University Press.

Noszczyk, T. (2019). A review of approaches to land use changes modeling. *Human and Ecological Risk Assessment: An International Journal 25*(6), 1377–1405.

Orcutt, G. H. (1957). A new type of socio-economic system. *The Review of Economics and Statistics 39*(2), 116–123.

Oren, T., M. S. Elzas, I. Smit, and L. G. Birta (2002). Code of professional ethics for simulationists. In *Summer Computer Simulation Conference*, pp. 434–435. Society for Computer Simulation International; 1998.

Paulus, J. K. and D. M. Kent (2020). Predictably unequal: understanding and addressing concerns that algorithmic clinical prediction may increase health disparities. *NPJ Digital Medicine 3*(1), 99.

Polhill, G., M. Borit, C. Elsenbroich, H. Verhagen, and N. Wijermans (2025). Ethical dimensions to empirical applications of agent-based social simulation. In *Advances in Social Simulation 2025*, pp. 411–424. Springer.

Poundstone, W. (1993). *Prisoner's Dilemma.* New York, NY: Anchor Books.

R Core Team (2021). *R: A Language and Environment for Statistical Computing.* Vienna, Austria: R Foundation for Statistical Computing.

Rademacher, C., C. Neuert, V. Grundmann, C. Wissel, and V. Grimm (2004). Reconstructing spatiotemporal dynamics of central european natural beech forests: the rule-based forest model before. *Forest Ecology and Management 194*(1–3), 349–368.

Railsback, S. F. and V. Grimm (2011). *Agent-based and Individual-based Modeling: A Practical Introduction.* New Jersey: Princeton University Press.

Richiardi, M., D. Collado, and D. Popova (2021). UKMOD–a new tax-benefit model for the four nations of the UK. *International Journal of Microsimulation 14*(1), 92–101.

Rogers, E. M. (2003). *Diffusion of Innovations* (5th ed.). Simon & Schuster.

Rosés, R., C. Kadar, and N. Malleson (2021). A data-driven agent-based simulation to predict crime patterns in an urban environment. *Computers, Environment and Urban Systems 89*, 101660.

Sage, L. and A. Flache (2021). Can ethnic tolerance curb self-reinforcing school segregation? a theoretical agent based model. *Journal of Artificial Societies and Social Simulation 24*(2), 2.

Sakoda, J. (1949). *Minidoka: An Analysis of Changing Patterns of Social Behavior.* Ph. D. thesis, University of California.

Sakoda, J. M. (1971). The checkerboard model of social interaction. *The Journal of Mathematical Sociology 1*(1), 119–132.

Salonen, J., H. Tikanmäki, and S. Lappo (2021). Partition of the life course: An extended dynamic microsimulation analysis. *International Journal of Microsimulation 14*(3), 54–75.

Sandholm, W. H. (2020). *Evolutionary game theory*, pp. 573–608. Springer.

Schelling, T. C. (1971). Dynamic models of segregation. *The Journal of Mathematical Sociology 1*(2), 143–186.

Schlüter, M., A. Baeza, G. Dressler, K. Frank, J. Groeneveld, W. Jager, M. A. Janssen, R. R.

McAllister, B. Müller, K. Orach, et al. (2017). A framework for mapping and comparing behavioural theories in models of social-ecological systems. *Ecological Economics 131*, 21–35.

Shults, F. L. and W. J. Wildman (2019). *Ethics, computer simulation, and the future of humanity*, pp. 21–40. Springer.

Silver, N. (2012). *The Signal and the Noise: The art and science of prediction.* Penguin UK.

Smith, J. M. (1964). Group selection and kin selection. *Nature 201*(4924), 1145.

Smith, P. L. and D. R. Little (2018). Small is beautiful: In defense of the small-n design. *Psychonomic Bulletin & Review 25*(6), 2083–2101.

Squazzoni, F., J. G. Polhill, B. Edmonds, P. Ahrweiler, P. Antosz, G. Scholz, É. Chappin, M. Borit, H. Verhagen, F. Giardini, et al. (2020). Computational models that matter during a global pandemic outbreak: A call to action. *Journal of Artificial Societies and Social Simulation 23*(2), 10.

Srivastava, P. and N. Hopwood (2009). A practical iterative framework for qualitative data analysis. *International Journal of Qualitative Methods 8*(1), 76–84.

Stoica, V. and A. Flache (2014). From Schelling to schools: A comparison of a model of residential segregation with a model of school segregation. *Journal of Artificial Societies and Social Simulation 17*(1), 5.

Sutton, T., O. Dassau, and M.Sutton (2009). A gentle introduction to GIS. Technical report, QGIS.

Szangolies, L., M.-S. Rohwäder, H. Ahmed, F. Jahanmiri, A. Wagner, R. Souto-Veiga, V. Grimm, and C. Gallagher (2024). Visual ODD: A standardised visualisation illustrating the narrative of agent-based models. *Journal of Artificial Societies and Social Simulation 27*(4), 1.

Tesfatsion, L. (2017). Modeling economic systems as locally-constructive sequential games. *Journal of Economic Methodology 24*(4), 384–409.

Tesfatsion, L. (2023). *Agent-based computational economics: Overview and brief history*, pp. 41–58. Springer.

Thaler, J. and P.-O. Siebers (2020). Specification testing of agent-based simulation using property-based testing. *Autonomous Agents and Multi-Agent Systems 34*(47).

Tolk, A. (2017). *Code of Ethics*, pp. 35–52. Wiley Online Library.

Troost, C., R. Huber, A. R. Bell, H. van Delden, T. Filatova, Q. B. Le, M. Lippe, L. Niamir, J. G. Polhill, Z. Sun, and T. Berger (2023). How to keep it adequate: A protocol for ensuring validity in agent-based simulation. *Environmental Modelling & Software 159*, 105559.

Ugwudike, P. (2022). Predictive algorithms in justice systems and the limits of tech-reformism. *International Journal for Crime, Justice and Social Democracy 11*(1), 85–99.

Verhagen, H., C. Elsenbroich, and N. Wijermans (2023). Agent decision-making heterogeneity—agent (meta) frameworks for agent-based modelling. In *Conference of the European Social Simulation Association*, pp. 621–630. Springer.

Volterra, V. (1926). Variazioni e fluttuazioni del numero d'individui in specie animali conviventi. *Memoria della Reale Accademia Nazionale dei Lincei 2*, 31–113.

Vu, T. M., C. Probst, J. M. Epstein, A. Brennan, M. Strong, and R. C. Purshouse (2019). Toward inverse generative social science using multi-objective genetic programming. In *Proceedings of the Genetic and Evolutionary Computation Conference*, pp. 1356–1363.

Weaver, W. (1948). Science and Complexity. *American Scientist 36*, 536–544.

Wijermans, N., G. Scholz, É. Chappin, A. Heppenstall, T. Filatova, J. G. Polhill, C. Semeniuk, and F. Stöppler (2023). Agent decision-making: The elephant in the room-enabling the justification of decision model fit in social-ecological models. *Environmental Modelling & Software 170*, 105850.

Wilensky, U. (1997). NetLogo Segregation Model.

Wilensky, U. (1999a). NetLogo http://ccl.northwestern.edu/netlogo/. Technical report, Center for Connected Learning and Computer-Based Modeling, Northwestern University, Evanston, IL.

Wilensky, U. (1999b). NetLogo voting model. http://ccl.northwestern.edu/netlogo/models/voting. Technical report, Center for Connected Learning and Computer-Based Modeling, Northwestern University, Evanston, IL.

Wilensky, U. (2002). NetLogo Prisoner's Dilemma N-Person Iterated Model.

Wilensky, U. and W. Rand (2015). *An Introduction to Agent-Based Modeling: Modeling Natural, Social and Engineered complex Systems with NetLogo.* Cambridge, MA: MIT Press.

Wilke, C. O. (2019). *Fundamentals of Data Visualization.* Sebastopol, CA: O'Reilly Media.

Williams, T. G., S. D. Guikema, D. G. Brown, and A. Agrawal (2020). Assessing model equifinality for robust policy analysis in complex socio-environmental systems. *Environmental Modelling & Software 134*, 104831.

Zahle, J. (2023). Methodological Holism in the Social Sciences. In E. N. Zalta and U. Nodelman (Eds.), *The Stanford Encyclopedia of Philosophy* (Winter 2023 ed.). Metaphysics Research Lab, Stanford University.

NetLogo Language

List of all the NetLogo commands, reporters and other keywords presented in this book. See the NetLogo Dictionary for all available keywords, their definitions and short examples. Some keywords also have short versions, but such abbreviations are not used in the tutorial. The full words are easier to read and more natural to remember.

Index

For Product Safety Concerns and Information please contact our EU representative GPSR@taylorandfrancis.com Taylor & Francis Verlag GmbH, Kaufingerstraße 24, 80331 München, Germany

Batch number: 10733596

Printed by Printforce, the Netherlands